WELL SERVICING

Artificial Lift Methods

Second Edition

By William Lane

originally produced in cooperation with

INTERNATIONAL ASSOCIATION OF DRILLING CONTRACTORS (IADC)
Houston, Texas

published by

PETROLEUM EXTENSION SERVICE
Division of Continuing & Innovative Education
THE UNIVERSITY OF TEXAS AT AUSTIN

2013

Library of Congress Cataloging-in-Publication Data

Lane, William, 1951–
Artificial lift methods / by William Lane. — Second edition.
pages cm. — (Lessons in well servicing and workover ; Lesson 5)
Includes bibliographical references and index.
"Catalog no. 3.70520."
ISBN 978-0-88698-258-4 — ISBN 0-88698-258-8 1. Secondary recovery of oil. 2. Oil wells—Artificial lift. I. Title.
TN871.37.L36 2012
622'.33827—dc23

2012023605

DISCLAIMER

Although all reasonable care has been taken in preparing this publication, the authors, the Petroleum Extension Service (PETEX) of The University of Texas at Austin, and any other individuals and their affiliated groups involved in preparing this content assume no responsibility for the consequences of its use. Each recipient should ensure he or she is properly trained and informed about the unique policies and practices regarding application of the information contained herein. Any recommendations, descriptions, and methods in this book are presented solely for educational purposes.

First Edition published 1971. Second Edition 2013.
Printed in the United States of America

Editors: Christopher Parker and Dewey Badeaux
Graphic Designer: Debbie Caples
Cover Art: E. K. Weaver

Catalog No. 3.70520
ISBN 978-0-88698-258-4
0-88698-258-8

No state tax funds were used to publish this book.
The University of Texas at Austin is an equal opportunity employer.

Contents

Figures

Tables

Foreword

The Committee for Education and Training of the American Association of Oilwell Drilling Contractors is the original sponsor of twelve lessons of PETEX™ training material concerning well servicing and workover. These lessons can be covered in roughly two years of home study time.

Most of the learning to be accomplished by those actually working in the servicing and workover industry will take place on the job with the aid of their supervisors and associates in the field. The purpose of these lessons is to help the new worker find out what the work involves, to better understand each role in the operation, and to expedite comprehension of the overall situation.

Lesson 5 discusses the means of supplementing reservoir energy to raise the fluid to the surface after a well ceases to flow. There are hundreds of thousands of oilwells in the United States, the vast majority of which are produced by artificial lift. Making artificial lift as efficient and reliable as possible is a hugely important undertaking. PETEX is committed to making the study of lift technologies as rewarding as possible to the user of this lesson.

Preface

The global demand for energy continues to rise in the foreseeable future. Alternative energy sources and efforts to conserve energy provide some relief, but they do not keep pace with the growing demand for energy. It remains up to the oil and gas industry to supply the majority of energy to the world.

There are two ways to meet the rising demand for hydrocarbon-based energy. First, we can find and produce hydrocarbons from new reservoirs not previously exploited. Production of new reservoirs will continue to become more challenging because most of the easy-to-reach hydrocarbons are already being produced. New hydrocarbon reservoirs will be deeper, and produced fluids will be hotter, more viscous, and chemically more challenging. Second, we must produce hydrocarbons from existing reservoirs more efficiently. In both cases, effective lifting of hydrocarbons to the surface will be critical to meet these production challenges.

Over 90% of active oil and gas wells depend on some type of artificial lift, so lift technologies must keep pace with the demands for more hydrocarbons. The performance envelopes of traditional lift technologies will continue to be expanded. New lift technologies will need to be developed to extract hydrocarbons from challenging reservoirs and to reach hydrocarbons previously left behind.

The content of this publication provides an overview of lift technologies as a starting point for understanding artificial lift. The material focuses more on concepts rather than performance envelopes, because those technologies continue to advance. Even as new lift technologies are developed, the concepts presented in this book will remain the foundation for the lift industry and the keys to providing energy for future generations.

William Lane
Vice President of Emerging Technologies
Weatherford Artificial Lift Systems

Acknowledgments

The main goal of this book, as with the others of the Well Servicing and Workover Series, is to help workers gain a basic knowledge of the process, equipment, and problems they might encounter in the field. The book also provides an overview of the subject to nontechnical staff who need to understand the significant issues involved with artificial lift.

PETEX is extremely grateful to William Lane and other artificial lift professionals at Weatherford International for their contributions to this book. Thanks go to Weatherford International for releasing the materials to PETEX, Shauna Noonan for her editing and contributions to the content, and ConocoPhillips for Shauna's time on the project.

We sincerely thank PETEX staff members E.K. Weaver and Debbie Caples for their work on the figures and design of this book. Additionally, we would like to thank Leah Lehmann for her invaluable proofreading.

Chris Parker and Dewey Badeaux, Editors
Petroleum Extension Service (PETEX™)
THE UNIVERSITY OF TEXAS AT AUSTIN

About the Author

William Lane has 35 years of experience in the oil and gas industry performing roles in engineering, manufacturing, global product line management, and artificial lift training. He has been directly involved with surface service equipment, completions, compression, artificial lift, and unconventional resources. He has been working with Weatherford International and the former EVI Oil Tools Ltd. for 18 years in various executive positions and is currently serving as the vice president of emerging technologies for Weatherford Artificial Lift Systems Inc.

Lane holds several U.S. patents, and in 2003 was the recipient of a Harts E&P Special Meritorious Award for Engineering Innovation. He holds a B.S. degree in Mechanical Engineering and an M.S. degree in Mechanical Engineering Design, both from the University of Texas at Arlington.

Units of Measurement

Throughout the world, two systems of measurement dominate: the English system and the metric system. Today, the United States is one of only a few countries that employ the English system.

The English system uses the pound as the unit of weight, the foot as the unit of length, and the gallon as the unit of capacity. In the English system, for example, 1 foot equals 12 inches, 1 yard equals 36 inches, and 1 mile equals 5,280 feet or 1,760 yards.

The metric system uses the gram as the unit of weight, the metre as the unit of length, and the litre as the unit of capacity. In the metric system, 1 metre equals 10 decimetres, 100 centimetres, or 1,000 millimetres. A kilometre equals 1,000 metres. The metric system, unlike the English system, uses a base of 10; thus, it is easy to convert from one unit to another. To convert from one unit to another in the English system, you must memorize or look up the values.

In the late 1970s, the Eleventh General Conference on Weights and Measures described and adopted the Systeme International (SI) d'Unites. Conference participants based the SI system on the metric system and designed it as an international standard of measurement.

The Well Servicing and Workover Series gives both English and SI units. And because the SI system employs the British spelling of many of the terms, the book follows those spelling rules as well. The unit of length, for example, is metre, not meter. (Note, however, that the unit of weight is gram, not gramme.)

To aid U.S. readers in making and understanding the conversion system, we include the table on the next page.

English-Units-to-SI-Units Conversion Factors

Quantity or Property	English Units	Multiply English Units By	To Obtain These SI Units
Length, depth, or height	inches (in.)	25.4	millimetres (mm)
		2.54	centimetres (cm)
	feet (ft)	0.3048	metres (m)
	yards (yd)	0.9144	metres (m)
	miles (mi)	1609.344	metres (m)
		1.61	kilometres (km)
Hole and pipe diameters, bit size	inches (in.)	25.4	millimetres (mm)
Drilling rate	feet per hour (ft/h)	0.3048	metres per hour (m/h)
Weight on bit	pounds (lb)	0.445	decanewtons (dN)
Nozzle size	32nds of an inch	0.8	millimetres (mm)
Volume	barrels (bbl)	0.159	cubic metres (m^3)
		159	litres (L)
	gallons per stroke (gal/stroke)	0.00379	cubic metres per stroke (m^3/stroke)
	ounces (oz)	29.57	millilitres (mL)
	cubic inches ($in.^3$)	16.387	cubic centimetres (cm^3)
	cubic feet (ft^3)	28.3169	litres (L)
		0.0283	cubic metres (m^3)
	quarts (qt)	0.9464	litres (L)
	gallons (gal)	3.7854	litres (L)
	gallons (gal)	0.00379	cubic metres (m^3)
	pounds per barrel (lb/bbl)	2.895	kilograms per cubic metre (kg/m^3)
	barrels per ton (bbl/tn)	0.175	cubic metres per tonne (m^3/t)
Pump output and flow rate	gallons per minute (gpm)	0.00379	cubic metres per minute (m^3/min)
	gallons per hour (gph)	0.00379	cubic metres per hour (m^3/h)
	barrels per stroke (bbl/stroke)	0.159	cubic metres per stroke (m^3/stroke)
	barrels per minute (bbl/min)	0.159	cubic metres per minute (m^3/min)
Pressure	pounds per square inch (psi)	6.895	kilopascals (kPa)
		0.006895	megapascals (MPa)
Temperature	degrees Fahrenheit (°F)	$\frac{°F - 32}{1.8}$	degrees Celsius (°C)
Mass (weight)	ounces (oz)	28.35	grams (g)
	pounds (lb)	453.59	grams (g)
		0.4536	kilograms (kg)
	tons (tn)	0.9072	tonnes (t)
	pounds per foot (lb/ft)	1.488	kilograms per metre (kg/m)
Mud weight	pounds per gallon (ppg)	119.82	kilograms per cubic metre (kg/m^3)
	pounds per cubic foot (lb/ft^3)	16.0	kilograms per cubic metre (kg/m^3)
Pressure gradient	pounds per square inch per foot (psi/ft)	22.621	kilopascals per metre (kPa/m)
Funnel viscosity	seconds per quart (s/qt)	1.057	seconds per litre (s/L)
Yield point	pounds per 100 square feet (lb/100 ft^2)	0.48	pascals (Pa)
Gel strength	pounds per 100 square feet (lb/100 ft^2)	0.48	pascals (Pa)
Filter cake thickness	32nds of an inch	0.8	millimetres (mm)
Power	horsepower (hp)	0.75	kilowatts (kW)
Area	square inches ($in.^2$)	6.45	square centimetres (cm^2)
	square feet (ft^2)	0.0929	square metres (m^2)
	square yards (yd^2)	0.8361	square metres (m^2)
	square miles (mi^2)	2.59	square kilometres (km^2)
	acre (ac)	0.40	hectare (ha)
Drilling line wear	ton-miles (tn•mi)	14.317	megajoules (MJ)
		1.459	tonne-kilometres (t•km)
Torque	foot-pounds (ft•lb)	1.3558	newton metres (N•m)

Beam-lift units at work

Artificial Lift Overview

In this chapter:

- How artificial-lift systems produce fluids
- Current lift technologies used on land and offshore
- Factors to consider when selecting a lift system
- Environmental and operator-discretionary factors

Ideally, a hydrocarbon-bearing reservoir should contain enough natural pressure to enable *fluids* to flow to the surface for several years without requiring external energy. Over time, however, energy in the formation will decline to the point that pressure and/or flow velocity will no longer be adequate to move fluids to the surface. When a well reaches this point in its lifecycle, fluids must be produced (or lifted) to the surface through artificial means.

Notable exceptions include wells completed in prolific *water drive* reservoirs where wells continue to flow water under natural reservoir energy after hydrocarbon production has ceased. Likewise, large *gas-cap reservoirs* can contain sufficient energy to produce much of the recoverable hydrocarbons without artificial lift. However, more often, wells require artificial lift at some point in their economic life.

Even gas wells typically require some sort of deliquification system to remove water. Water accumulating in the wellbore creates a back-pressure that limits gas inflow from the reservoir, so the water must continually or periodically be removed to allow for the free flow of gas.

Types of Lift Systems

Most artificial-lift systems can be grouped into the following categories based on the type of energy used to lift the well fluids and the way that the energy is used. The characteristics and implications of each pumping technology are discussed in more detail in subsequent sections.

Reciprocating Rod Lift

Reciprocating rod lift, or *sucker rod pumping*, is by far the most commonly used method of artificial lift. It consists of a subsurface reciprocating rod pump that is connected to a surface pumping unit by a length of *sucker rods* or *continuous rod*, known as rod string. The surface pumping unit produces an up-and-down motion in the rod string, which strokes a *piston* in the subsurface pump. The reciprocating piston lifts the fluids on the upstroke and falls back through the static fluid column on the downstroke.

Electric Submersible Pumps

Electric submersible pumps (*ESPs*) consist of a subsurface electric motor that powers a *multistage centrifugal pump*. Electricity is transmitted through a cable to a subsurface motor. ESP assemblies include seal sections that keep well fluids out of the motor. The seal sections provide a means of balancing or compensating between interior and exterior pressures. Seal sections also transfer *torque* from the motor to the pump and handle axial thrust generated by the pump and motor components.

Progressing Cavity Pumps

Progressing cavity pumps (*PC pumps* or *PCPs*) consist of a subsurface rotary pump typically connected to a surface drivehead by means of a rotating rod string. Although sometimes referred to as *screw pumps*, PC pumps are *positive-displacement pumps* that move fluids through sealed cavities toward an outlet as a rotor turns inside a stator. PC pumps can also be driven by subsurface, ESP-style electric motors using a speed reducing gearbox or special low-speed, high-torque motors. These bottom-driven PC pumps are sometimes referred to as *electric submersible progressing cavity pumps* (*ESPCPs*).

Conventional Gas Lift

Conventional *gas-lift* technologies are classified as either *continuous* or *intermittent gas lift*.

Continuous gas lift uses injected gas to lighten the weight of the fluid column inside the wellbore so fluids can reach the surface.

Typically, compressed gas from the surface travels down the annulus between the casing and the production tubing. This gas enters the production tubing through valves located inside *mandrels* in the production tubing. The resulting bubbles reduce the average density of the fluid above the injection point. When the *hydrostatic pressure* (*HP*) in the fluid column becomes less than the flowing *bottomhole pressure* in the reservoir, the fluid column will flow to the surface.

Intermittent gas lift uses the velocity of compressed gas to displace well fluids to the surface. With this technology, gas is periodically released suddenly from the well annulus into the production tubing. The resulting high velocity gas displaces slugs of well fluids to the surface.

Plunger Lift

Plunger lift is similar to intermittent gas lift in that it periodically removes slugs of liquid rather than lifting the entire column of fluid. With this method, a *plunger* is released from the surface and falls through the standing liquid level in the well. For conventional plunger systems, the well is temporarily shut-in to allow formation gas to accumulate under the plunger. The surface flow line is then opened, causing a pressure drop across the plunger. The higher pressure gas below the plunger propels the plunger to the surface. This displaces the liquid slug out of the well and into the flow line. At the surface, the plunger can be caught and held in a special lubricator until a timer releases it for its next lift cycle. Continuous plunger systems operate while the well is flowing. A bypass valve in the plunger allows it to fall against production flow. When the plunger reaches the downhole assembly, the bypass valve closes, allowing the well pressure to lift the plunger and any accumulated liquids above the plunger to the surface.

Velocity Strings and Foam Lift

In gas wells, the gas flow can be used to remove liquids so that the liquids do not inhibit gas production. *Velocity strings* can be used to increase gas flow velocities above *critical velocity* to displace liquids to the surface. Gas injection can also be used to keep gas velocities elevated. Chemical *surfactants* can be added to reduce the surface tension of liquids, which causes liquids to foam when exposed to gas turbulence. The liquids become disbursed as film within the foam bubble matrix. The resulting foamy fluid is more easily carried to the surface by the gas stream. Foam lift has been used primarily to remove water from gas wells, but surfactants have also been developed to effectively foam hydrocarbon liquids.

Hydraulic Lift

Hydraulic lift makes use of two pumping technologies to bring well fluids to the surface.

Hydraulic jet pumps use high-pressure *power fluid* (usually water) from the surface to produce well fluids through a subsurface venturi-type injector nozzle. The power fluid exits the *venturi nozzle* at high velocity, creating a low-pressure area in the venturi inlet to the surrounding production fluids. The production fluids are drawn into the power fluid flow stream where they comingle with the power fluid flowing back to the surface.

Hydraulic piston pumps use high-pressure power fluid from the surface to stroke a subsurface pump piston in a power cylinder, which in turn strokes a plunger in a subsurface *reciprocating pump*. At the end of the power stroke, the power fluid is automatically diverted to the other side of the power cylinder, reversing the motion so the pump reciprocates automatically in the presence of a constant supply of power fluid.

Lift System Selection

It is important to gather information about well construction, fluid composition and properties, location and infrastructure, and any historical lift data (about the well and nearby wells) prior to selecting an artificial-lift system. Once data is collected and understood, artificial-lift selection involves a three-step process.

A number of items should be reviewed prior to selecting an artificial-lift system, including historical data and information about nearby wells.

Step 1. Determine Target Production Conditions

The target lifting pressure that the pump must generate will be the sum of the hydrostatic pressure (HP) created by the weight of the fluid column in the well plus friction losses resulting from flow through the tubing and any back-pressure caused by flow lines and surface equipment. This pressure is often expressed in equivalent feet of water[1] to more simply relate the application depth to the system performance.

The target production capacity for the lift system should be the expected production fluid rate from the reservoir plus any injected fluids. Because most well fluids contain some gas, the total volumetric rate to be lifted must include gas volumes at flowing pressures. Also, lift systems should be sized to anticipate changes in lift volumes.

[1] 1 foot of water at standard conditions = 0.4335 psi.

Initially, production rates might decline with time; however, injected fluids from waterfloods and similar enhanced production practices can increase the fluid volumes to be lifted. Over time, water migration within the reservoir can also impact the fluid volume to be lifted. It is good practice to size pumping systems conservatively according to the uncertainty of future lift requirements.

Step 2. Eliminate Unviable Lift Technologies

With the target production conditions known, the second step in the selection process is to eliminate all lift technologies that are not capable of satisfying the target production conditions. Table 1 shows the more common types of lift technologies available, their operating limits, and basic application information necessary to match technologies to applications. This is just a starting point, as the technologies not yet eliminated must then be examined in more detail to determine whether or not they can meet the performance requirements of the application. For example, rod pumps can lift fluid from depths of 16,000 feet (4,877 metres) and can produce volumes of 6,000 barrels per day but not concurrently. Limitations in rod strength and other factors limit the volume that can be lifted from deep wells. Similarly, lifting at high rates can be limited to relatively shallow well applications.

Step 3. Identify Technologies Most Suited for Production

The final step in the selection process is to then choose from the remaining viable lift methods based on economics and the decision factors for artificial lift discussed below.

There are five general categories of factors to consider when choosing an artificial lift method:

- Reservoir factors
- Well and flow line factors
- Fluid factors
- Environmental and regulatory factors
- Operator-specific discretionary factors

Table 1
ARTIFICIAL LIFT TECHNOLOGY COMPARISON

Reciprocating Rod Lift

Advantages	Challenges	Typical Applications
Most widely used and understood method	Visually obtrusive	Land vertical wells
Temperature and chemical tolerance	Rods limit offshore use (SSSV)	CBM dewatering
Rugged	Rod couplings can wear tubing	Thermal applications
Some configurations tolerate sand and gas		
High system efficiency		
High salvage value for surface pumping units		

Electric Submersible Pumps

Advantages	Challenges	Typical Applications
High volume and depth capability	Must have electricity	Offshore or land
Effective in deviated wells	Intolerant of sand and viscous liquids	High volume pumping
Small surface footprint	Efficiency is impacted by free gas	Medium and low viscosity
Low profile		CBM dewatering

Progressing Cavity Pumps

Advantages	Challenges	Typical Applications
Pumps viscous fluids	Rotor/stator fit must be tuned to each application	Cold heavy oil land
Particulate matter tolerance	Temperature limited	CBM dewatering
Highest system efficiency	Sensitive to CO_2 gas and aromatic gases	
Low profile	Sensitive to some well treatment fluids	
Small surface footprint	Rods limit offshore use (SSSV)	
	Rod couplings can wear tubing	

Conventional Gas Lift

Continuous Gas Lift

Advantages	Challenges	Typical Applications
Uses well energy	Significant surface infrastructure	Offshore or land
High flow rate capacity	Less effective in heavy/viscous oil	Multi-well projects
Sand tolerant	Limited well drawdown capacity	Where injection gas is present
Tolerant of high gas volumes	Requires sufficient formation pressure	
GL valves run on wireline	Sensitive to flow-line, back-pressure	
Centralized surface equipment		

Intermittent Gas Lift

Advantages	Challenges	Typical Applications
Uses well energy	Limited volume <500 BPD[2]	Low volume production
Sand tolerant	Limited well drawdown capacity	Deep production
Not impacted by gas	Flow surges on surface equipment	Low to moderate PI
Good in deviated wells	Sensitive to flow-line, back-pressure	
Can produce very low volumes		
Lower drawdown than conventional GL		
Surface equipment can be centralized		
GL valves can be run on wireline		

[2] Barrels per day

Table 1, cont.
ARTIFICIAL LIFT TECHNOLOGY COMPARISON

Plunger Lift

Advantages	Challenges	Typical Applications
Lowest cost lift method	Gas-liquid ratio requirements	Low volume stripper wells
Uses well energy	Volume limited <200 BPD	Unloading wells
No outside power required	Poor solids handling	Gas well dewatering
Simple installation (no rig)	Monitoring required to optimize	Offshore or land
Easy maintenance	Sensitive to flow-line, back-pressure	High GOR wells
Not impacted by gas		Wells with paraffin
Good in deviated wells		
Small surface footprint		
Can produce very low volumes		
Can scrape paraffin		

Velocity Strings and Foam Lift

Advantages	Challenges	Typical Applications
Inexpensive	Gas-liquid ratio requirements	Dewatering (including laterals)
Simple installation	Volume limit <500 BPD	
Uses well energy	Monitoring required to optimize	
Tolerates moderate solids	Limited well drawdown	
Small footprint	Sensitive to flow-line, back-pressure	
Capillary transport of chemical additives to remove and prevent deposits of scale, paraffin, salt, and corrosion		

Hydraulic Lift

Hydraulic Jet Pumps

Advantages	Challenges	Typical Applications
Most versatile lift technology	Low system efficiency	Land
No moving parts downhole	Large surface footprint	High production volumes
High volume capability		Deviated wells
Wireline or circulate to run and retrieve		Flowing back fracturing fluids
Deviated wells		Initial dewatering and kickoff
Tolerant of solids, corrosive fluids, and gas		

Hydraulic Piston Pumps

Advantages	Challenges	Typical Applications
Positive displacement	Intolerant of solids	Land
Run/retrieve via wireline or circulation	Intolerant of free gas	
Deviated wells		

Decision Factors

Reservoir Factors

Reservoirs with high fluid mobility (typically low-viscosity fluids in highly *permeable* formations) can maintain pressure at the wellbore easier than reservoirs with lower fluid mobility. When mobility is uncertain, low, or when flow rates are to be maximized by drawdown, lift designs should assume that the pump must have sufficient pressure capacity to lift the full column of fluid with little or no pump inlet pressure.

Water drive reservoirs tend to replenish and retain pressure during production more effectively than other reservoir drive mechanisms, although the oil percentage gives way to more water over time. Accordingly, the natural fluid height in the wellbore is often high enough to be compatible with gas lift as well as reducing the pressure requirements for all lift methods.

Gas drive wells rely on a fixed amount of gas to keep the well pressurized. This can be *free gas*, often accumulated in a gas cap if the reservoir pressure is below the gas bubble point, or the gas can be in solution if the reservoir pressure is above the bubble point. Because the gas does not replenish itself, the reservoir pressure will drop more quickly over time compared to water drive reservoirs, resulting in a decline in production rates. Artificial-lift systems should be selected with a sufficiently broad operating range (turndown ratio) to accommodate anticipated changes in production rates.

Reservoirs that are to be produced with water flood, carbon dioxide (CO_2) flood, or similar injection-based *enhanced oil recovery* (*EOR*) methods can produce increasing volumes of fluid as injection rates are increased. Therefore, the lift systems must be sized to accommodate planned increases in production rates. Systems that must produce CO_2 will need to utilize materials resistant to related acids and *elastomers* that are resistant to CO_2 absorption.

Well and Flow Line Factors

The inflow characteristics of each well are dependent upon reservoir factors and factors related to well construction and completion. In general, as the pressure in the wellbore is lowered below the static reservoir pressure, the reservoir will produce fluids into the well in proportion to the pressure drop between the well and the static reservoir pressure. This relationship defines the *productivity index* (*PI*) of the well:

$$PI = \text{produced volume of fluid per day} / (\text{static reservoir pressure} - \text{flowing well pressure})$$

This linear relationship will approximate actual well performance for water drive wells in which formation fluids are relatively incompressible and receive energy from relatively constant sources. However, wells with gas in solution (solution drive wells) or that have a gas cap tend to have a nonlinear relation between production and pressure when below bubble-point pressure.

Inflow Performance Relationship

IPR is unique to each individual well, so artificial-lift requirements should be evaluated for each individual well.

Inflow Performance Relationship (IPR) describes the actual inflow response to pressure differentials. The IPR takes into account the fluid characteristics (compressibility, *viscosity*, and surface tension), reservoir permeability (the ability of the formation to allow fluid to flow), and the restrictions to flow around the wellbore caused by well construction and well treatments. Therefore, IPR is unique to each individual well, which implies that artificial-lift requirements are likewise unique to each individual well.

For new wells, the IPR of nearby wells will be a reasonable starting point for selecting the best lift system. In new reservoirs that do not have a history of recorded reservoir flow and pressure relationships, drill stem tests and electric logs can provide some predictive flow volume and pressure information to help with lift system selection. Where lifting equipment is required on the first well completed in a new field, a temporary lift system is recommended until the reservoir flow characteristics are understood. Most lift methods lend themselves to pilot test installations.

Fluid production generally declines over the life of a well as the formation pressure declines. The resulting *decline curve* (production versus time) implies that the lift requirements over the life of the well will change. Well operators should evaluate the economics of changing lift systems as lift requirements change, or select systems that can efficiently lift over a broad range of production volumes and flowing bottomhole pressures.

A well that is capable of producing a large volume of fluid in excess of 5,000 barrels of fluid per day would typically require a high-volume lift technology such as gas lift, electric submersible pumps, or hydraulic jet pumps. Pump flow rates are usually kept below the maximum possible flow rates in order to prevent formation damage and to keep the fluid level in the well sufficiently higher than the pump intake to avoid pump damage.

Well depth is the primary factor for determining the pressure that the pump must generate in order to lift the fluids from the pump landed depth to the surface. Other well factors that impact pump pressure rating are tubing size, fluid viscosity, flow restrictions, surface back-pressure, and flow rates versus IPR. Pump power requirements are determined by the differential pressure across the pump for a given flow rate and assumed pumping efficiency.

Well Configuration Factors

Offshore well designs typically require *subsurface safety valves (SSSVs)*, which are incompatible with rod-driven artificial-lift systems. However, deep-set SSSVs placed below the lift system can be used if regulations allow them. Most offshore artificial lift is gas lift or electric submersible pump lift. Hydraulic pumps can be used with subsurface safety valves but are generally not used offshore due to surface equipment size and weight.

Well *deviation* presents another challenge for conventional rod-driven systems. Rod coupling wear on the production tubing worsens due to increased side loads in curved sections of the tubing, although there are ways to minimize this type of wear. Reciprocating rod systems have the additional challenge of slower rod fall rates in deviated sections since gravity is not acting in the same direction that the rods must fall. Hydraulic pumps are ideal for deviated wells because they can be pumped down from the surface to the subsurface pump location and retrieved to the surface by reverse-circulating the well fluids. Electric submersible pumps are also effective in deviated wells. While gas lift can also be used, its lift efficiency decreases as deviation increases because the lift gas tends to separate out of the fluid column due to natural gravity segregation.

Flow-line pressure increases the required lifting pressure in any artificial-lift application, but it is particularly challenging to lift systems in applications with low reservoir pressures. Some of the more economical and low-volume systems will be excluded from consideration if flow-line pressures are too high. Flow-line pressures should be minimized to the extent practical within the field gathering system.

Fluid Factors

Produced gases can impact pump efficiency, and some gases such as CO_2 can be detrimental to elastomers in subsurface pumps.

When choosing lift equipment, it is important to consider the reservoir fluid that will be produced. For example, high *gas-liquid ratio (GLR)* fluids are challenging for many pump technologies because of gas compressibility, potential pump cavitation, reduced pump cooling, and reduced fluid lubricity. In particular, some pumps might *gas lock*—that is, gas can displace all of the liquid in the pump causing the pump cycle to compress and expand the gas without moving any fluids through the pump. *Gas interference* is similar to gas lock in that gas partially displaces liquid in a pump, causing it to move less fluid as the gas compresses and expands. Gases such as CO_2 and aromatic gases can be detrimental to elastomer products, so pumping systems with fewer elastomer components are often used in the presence of these gases. Many lift systems have special features to reduce the negative effects of gas.

Some crude oils contain *paraffin*, which can be deposited in the wellbore and on well equipment when oil pressure or temperature is reduced. High-paraffin oils complicate the use of gas lift due to paraffin formation from temperature drops caused by gas pressure drops across the *gas-lift valves*. Jet pumps can be effective in paraffinic wells if the power fluid is heated.

Scale and mineral deposits can likewise compromise lifting systems. Lift systems should have materials and coatings compatible with well fluids. Production chemicals can be injected to inhibit scale and mineral formation and deposition. Flow assurance and conformance refer to technologies that focus on preventing formation and deposition of paraffin, scale, and minerals.

High-viscosity fluids, including asphaltic oils, also present challenges. They are often found in unconsolidated sand formations, resulting in fluid that is sand-laden. Screening the sand from *viscous* oil can be difficult, so pumps used with viscous fluids should be sand-tolerant. Progressing cavity pumps are excellent at pumping viscous, sandy fluid at moderate temperatures. Heating and chemicals can be used to reduce viscosity, but the PCP elastomer might be compromised if not designed for the higher temperature and chemical exposure. The productivity index of the high-viscosity fluid reservoirs is usually low because viscous fluids flow less easily through formations. As a result, near pump-off conditions might be required at the pump in order to obtain a reasonable amount of daily oil production. Therefore, pumps in viscous fluids must be sized to lift a full column of fluid with little or no pump inlet pressure. Reciprocating rod pumps are more tolerant of pump-off conditions compared to other pumps, but pump speed in viscous fluids is limited by the speed that rods can fall through the thick fluids.

Environmental and Regulatory Factors

Protection of the environment is an influential factor in choosing an artificial lift method. Produced water and other well byproducts must be managed in order to prevent pollution. Water sourcing, treatment, and disposal should be part of the planning process for field development.

City and state regulations sometimes require that all producing equipment be hidden from sight or disguised. Visibility constraints favor the installation of low-profile systems rather than conventional reciprocating rod lift units.

The environment can be negatively impacted by well byproducts and poor lift maintenance. Effective lift management is important to reduce the risk of pollution.

Operator-Specific Discretionary Factors

Operators must consider factors that are unique to their company and operating environment. Availability of capital or a focus on short-term versus long-term profitability can constrain the selection of lift systems. One way to eliminate up-front capital expense for lift technologies is to arrange lease and rental contracts for equipment and services. Generally, the long-term cost of leasing will be higher than the amortized cost for procurement of equipment since the leaser must assume more risk.

Availability of labor, infrastructure, and other local resources can constrain lift selection. Some technologies are inherently more intuitive and easier to learn, so they are easier to introduce into locations with limited technical resources.

Familiarity with technologies is a common and valid selection criterion. Introduction of new technologies introduces risk of misapplication and early-cycle failures while personnel become familiar with the new technology. Inventory and logistics are likewise complicated as new technologies are introduced because a wider variety of support parts, supplies, and equipment must be managed. The result can be a real cost that will erode the expected benefits of introducing an unfamiliar technology.

Summary

Most wells will require artificial lift to produce oil or gas once natural reservoir energy drops past a certain point. The categories of lift systems include: reciprocating rod lift, conventional gas, plunger, velocity strings and foam, hydraulic jet, hydraulic piston, electric submersible, and progressing cavity pumps. For a given well, a lift system is selected by first defining the target production conditions and eliminating all lift technologies that are not capable of satisfying the target production conditions. Then the lift technology is chosen from the remaining viable lift methods based on economics and other decision factors.

Reciprocating Rod Lift

In this chapter:

- Typical applications of reciprocating rod systems
- Operating principles of a sucker rod pump
- Rod pump system design and components
- Types of surface rod pumping units
- Factors to consider when selecting and using rod string

The majority of artificial-lift systems in use are reciprocating rod lift systems.

The history of reciprocating rod lift is closely tied to the early oilwells that were established in 1859 by Edwin Drake in the small, rural community of Titusville, Pennsylvania. Commonly referred to as the Drake well, this earliest of drilling sites forever shaped industry and trade while advancing human mobility. Around 300 to 400 gallons (about 1,135 to 1,514 litres) were reportedly lifted from the site each day; however, the drilling process was expensive, tedious, and extremely dangerous.

Within ten years of the Drake well, conventional rod pumping was becoming increasingly popular. Early rod-pumping systems consisted of a standard cable tool drilling rig, placed in such a way that the walking beam could be used to operate the pump. Prior, rod-activated pumps had been used to produce *brine*. Similar to the pump illustrated in figure 1, they consisted primarily of a cylinder made up in the tubing string, a *standing valve* seated in the tubing string, a plunger, and *traveling valve*. It is likely that *flapper valves* were used rather than *ball valves*, which are depicted in the figure. Originally, the plunger was reciprocated in the cylinder by means of wooden sucker rods with wrought-iron end fittings for connections.

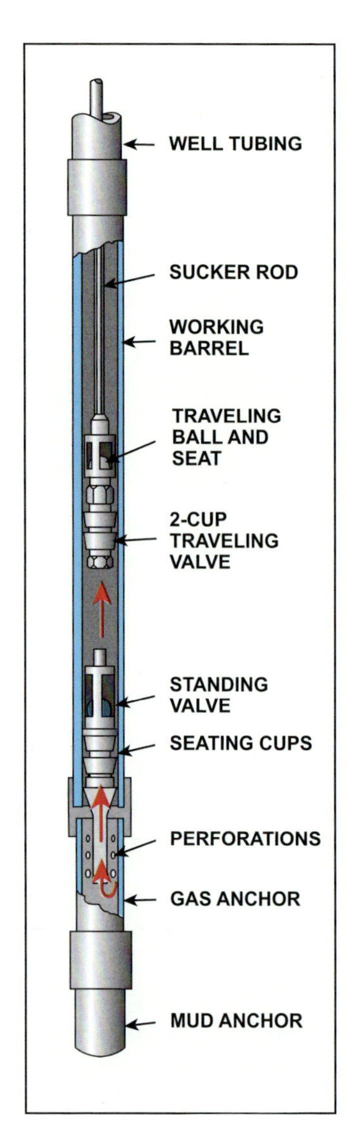

Figure 1. General arrangement of a reciprocating rod pump

Courtesy of Weatherford International

Figure 2. Rod pumping unit

One major improvement in reciprocating rod pump technology came with the development of the standard pumping rig, at first in a rather primitive form, but later in the more efficient design illustrated in figure 2. These rigs were first driven by steam power, subsequently by gas engines, and then by electric motors as depicted in figure 3. After electricity became generally available, it was often necessary to replace the engine with a countershaft as shown in figure 4 to obtain the speed reduction needed for an electric motor. Equipment continued to evolve to reach greater depths and to produce higher volumes of oil. Today, reciprocating rod lift systems are among the most efficient artificial-lift technologies and account for approximately 80% of all artificial-lift systems in operation.

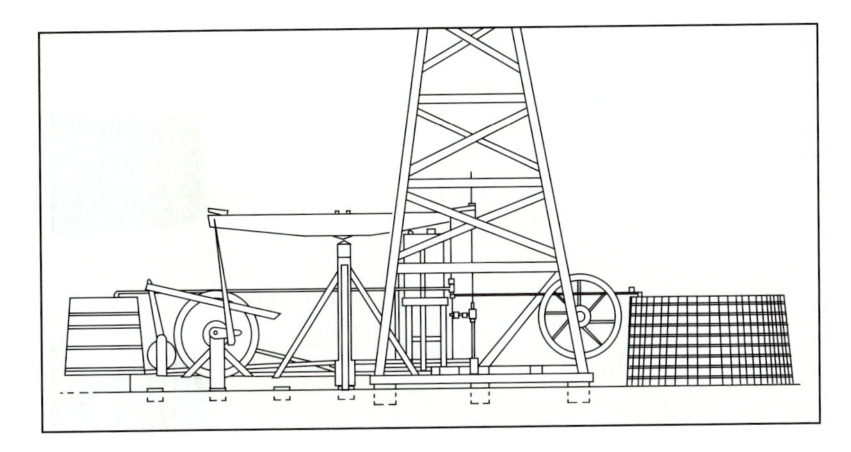

Figure 3. Early standard rig front arranged for pumping

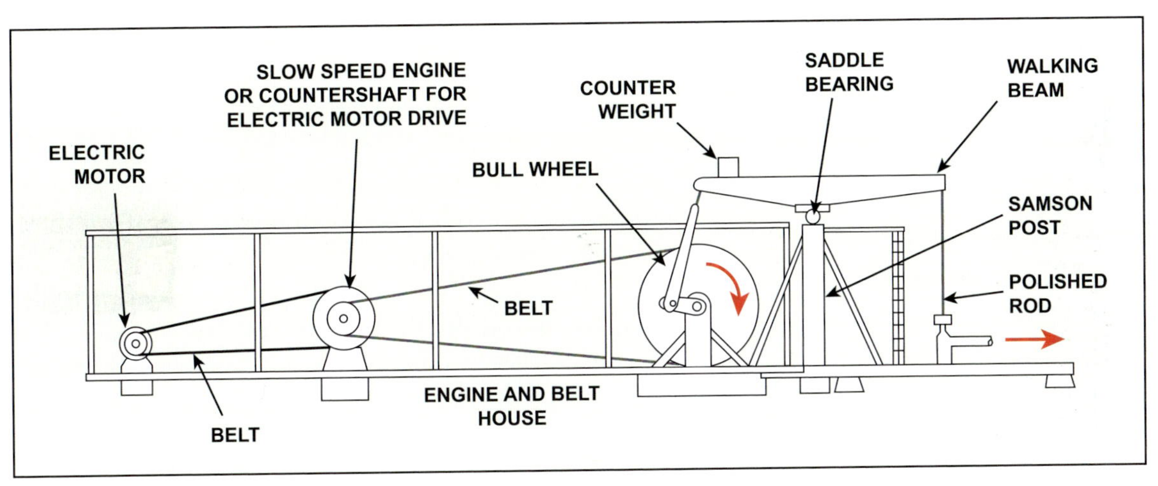

Figure 4. Early standard pumping rig driven by an electric motor drive

Typical Applications

Because reciprocating rod systems are cost-effective, easy to maintain, and highly productive, they are widely used for land applications.

Reciprocating rod systems are the most widely used lift method for land applications. Because they are simple, efficient, and require little attention, they remain the preferred lift system of many producers. The surface pumping units are rugged and have very long lives if properly maintained, so capital acquisition costs of pumping units can be amortized over a long period of time. This helps to make the pumping units more cost-effective in marginal applications. A broad range of sizes and configurations allow reciprocating rod systems to lift 5 to 5,000 barrels of liquid per day.

Subsurface reciprocating rod pumps can be used to lift most types of fluids. By adding weight to the rod string, reciprocating rod pumps can lift *heavy oil* (down to 8°API) that is over 100°F (38°C) or highly water-cut (in excess of 90%); the extra weight allows the rod strings to fall through the viscous oil on the downstroke. Rod pumps can be configured with no elastomeric parts for use in thermal applications.

Rod strings prevent subsurface safety valves from closing. For this reason, reciprocating rod systems are typically not used for offshore applications.

Tolerance to gas is fair with special configurations to help prevent gas lock. Tolerance to particulate matter is fair with the use of special sand-tolerant pumps, although typically rod pumps are used in wells with little or no particulate matter or in conjunction with screens.

The sucker rods are the primary limiting factor for reciprocating rod applications. In offshore applications, the rod string would prevent standard subsurface safety valves from closing. Even with deep-set subsurface safety valves, rod handling on offshore platforms is challenging. Thus, reciprocating lift is not typically considered for offshore use. Jointed sucker rod is not often used in applications with wells deviated over 20° per 100 feet (31 metres) because high side forces in the rods can cause the rod couplings to damage the production tubing. Centralizers on the rods can help prevent tubing wear, but they increase cost and must be replaced as they wear. Production tubing rotators can be used to slowly rotate the tubing within the casing to help spread out the tubing wear and thereby extend tubing life. Tubing wear can also be reduced significantly by using a continuous sucker rod. With continuous sucker rod strings, the side forces between the rod and tubing are spread out over a much greater contact area rather than being concentrated at the couplings. Continuous rod also has less flow resistance than coupled sucker rods, which makes it preferable for use in viscous fluids and in smaller tubing. The use of poly-lined tubing has reduced tubing wear in many fields.

The size of most pumping units makes them undesirable near many communities and in locations where natural beauty is a priority. Also, the prime movers, gear reduces, and large moving parts of pumping units tend to be loud. The use of pumping units, therefore, might be restricted in areas that are sensitive to noise. Special low-profile units are available for use under agricultural sprinkler systems.

Operating Principles

The subsurface sucker rod pump consists of a hollow plunger, which reciprocates within a cylinder, and two *check valves*: a traveling valve within the plunger assembly and a standing valve at the base of the pump as shown in figure 5. The pump cycle for a conventional tubing pump is shown in figure 6. On the upstroke, the weight of fluid in the tubing will cause the traveling valve to close. The upward motion displaces the fluid in the pump into the production tubing. As the plunger moves upward, it supports and moves the entire fluid column toward the surface. This upward motion of the plunger also causes a reduction of pressure within the pump barrel below the plunger.

The upward motion of the rod pump plunger lifts fluids to the surface.

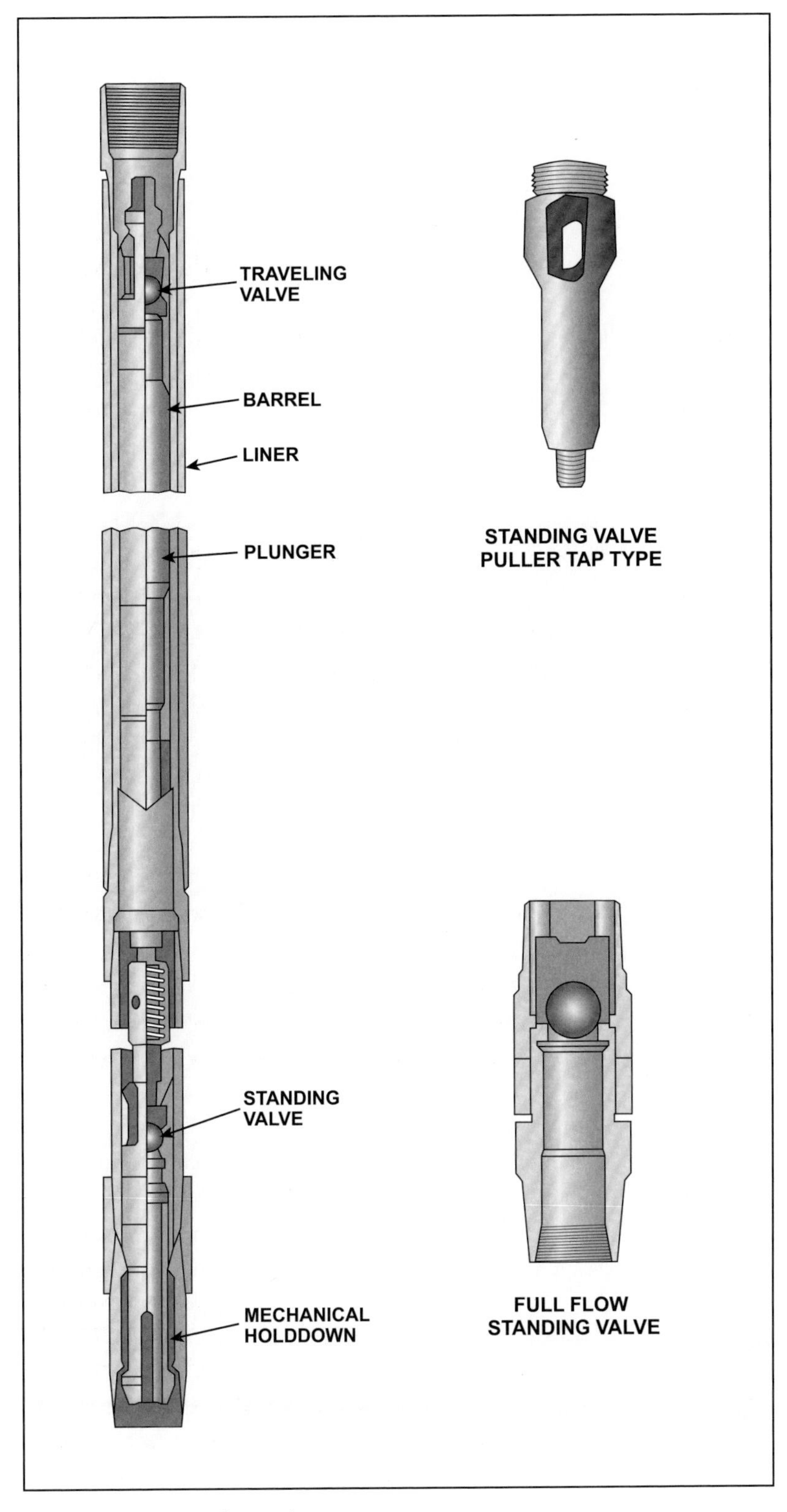

Figure 5. Subsurface rod pump

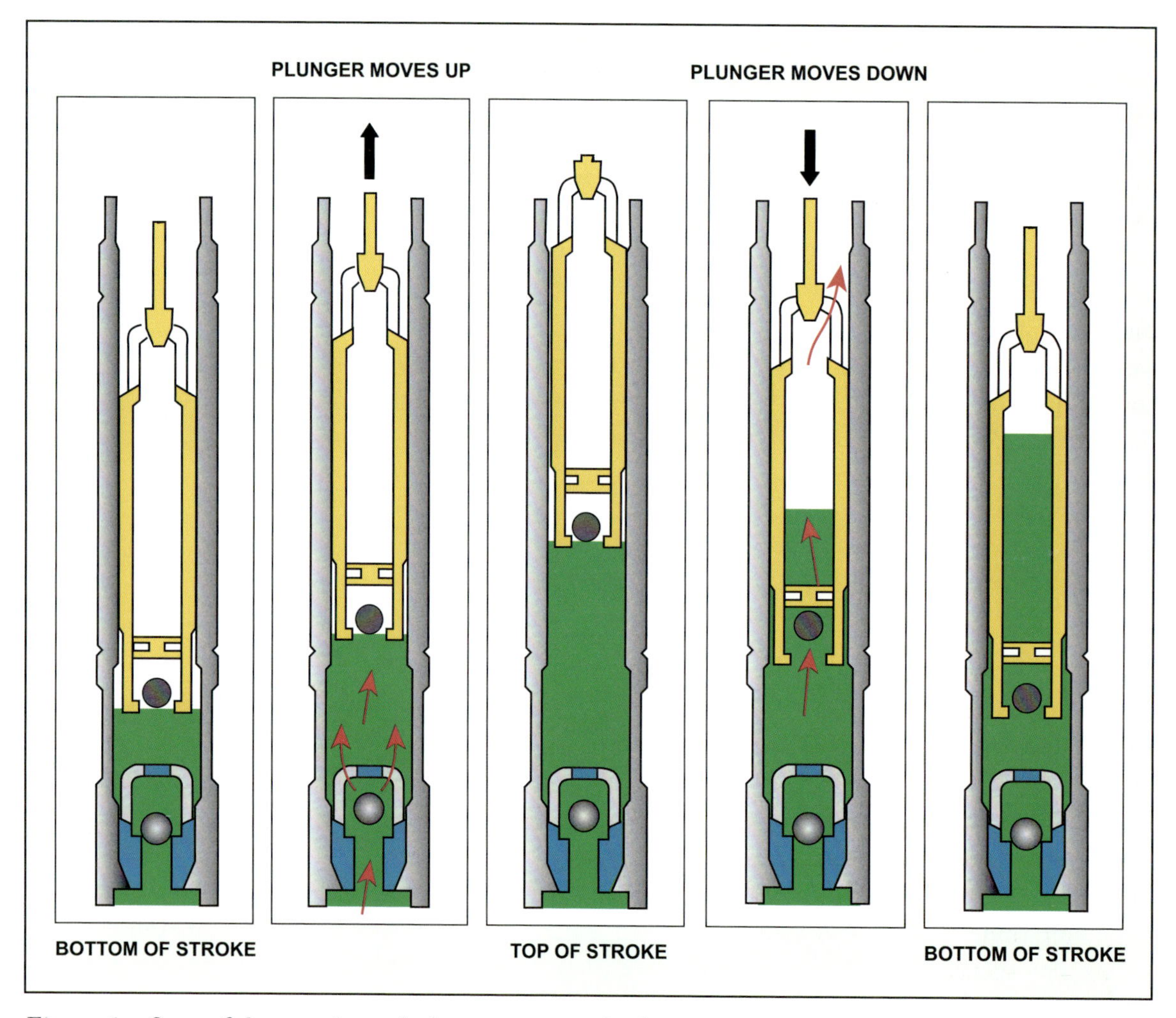

Figure 6. Steps of the pumping cycle for a conventional tubing pump

The pressure of the fluid in the well casing then opens the standing valve at the bottom of the pump barrel. The well fluid flows through the standing valve and fills the void left by the plunger on the upstroke.

On the downstroke, the downward motion of the plunger allows the fluid column weight to act on the standing valve, causing it to close. With the fluid trapped above the standing valve, further downward motion of the plunger causes the traveling valve to open. As the plunger continues downward, the fluid below the plunger passes through the traveling valve, filling the barrel above the plunger. In this way, the chamber above the plunger becomes filled and ready for the next upstroke. On the next upstroke, the plunger displaces the fluid from the pump into the tubing. Thereafter, the upstroke of each successive cycle raises this increment of fluid a given distance up the tubing until it finally reaches the surface and is produced into the lease tanks. On the downstroke, the plunger merely drops through the fluid that entered the pump through the standing valve on the preceding stroke.

During the upstroke, the weight of the fluid and the rods causes the rod string to stretch elastically. During the downstroke, the fluid weight is taken off of the rods so they contract. This stretching and contracting of the sucker rods causes plunger travel to typically be less than the stroke length of the surface pumping unit. At some pump rates, the dynamic response of the system can cause the plunger travel to exceed the stroke length of the surface pumping unit. This plunger travel (potentially *undertravel* or *overtravel*) must be considered in selecting pumping unit stroke length and pump length.

Pump Types

The majority of reciprocating rod systems use insert pumps because they are easier to retrieve and replace.

There are two basic types of rod pumps: the *tubing pump* and the *insert pump*. The tubing pump barrel is threaded directly onto the bottom of the production tubing and is installed with the tubing (fig. 7). The plunger is run into the tubing separately on the rod string. Generally, a standing valve puller is run on the bottom of the plunger. This allows the fluid in the production tubing to be drained prior to pulling the tubing, and it makes removal of the standing valve optional when pulling the sucker rods and plunger. The plunger, traveling valve, and standing valve can be serviced simply by pulling the rods, but the tubing must be pulled to service the pump barrel.

Insert pumps are run as complete units into the well inside of the production tubing on the rod string. The entire insert pump can be removed and serviced merely by pulling rods. For this reason, insert pumps are preferred over tubing pumps in most applications. A seating assembly or *hold-down* on the pump fits into the profile of a seating nipple in the tubing when the pump is run in. Because they must fit inside the production tubing, insert pumps have less volume capacity than comparable tubing pumps. Good production practice requires that the outside diameter of an insert pump be at least ¼-inch (6.35-millimetre) smaller than the nominal inside diameter of the tubing in order to be run through the tubing.

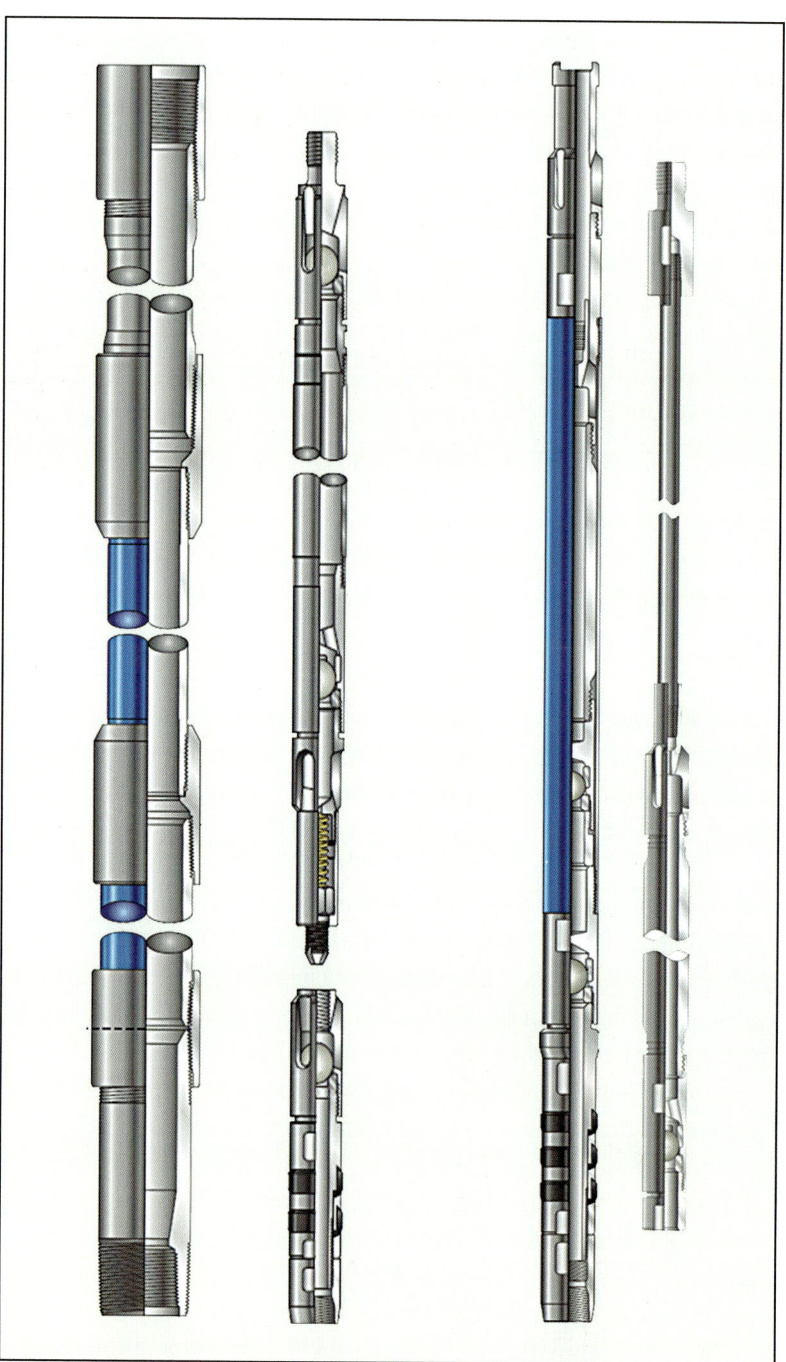

Courtesy of Weatherford International

Figure 7. Tubing pump and insert pump

Conventional stationary barrel insert pumps have a reciprocating plunger inside the barrel. Traveling barrel insert pumps utilize a stationary plunger anchored to the seating assembly and a barrel that reciprocates with the rod string motion. Traveling barrel pumps improve agitation to help keep particulate matter in solution but are not recommended in wells with significant gas production or in deep wells.

Among other characteristics, insert pumps can be classified by their anchor configuration as illustrated in figure 8:

- A stationary barrel top anchor pump has the hold-down at the top of the barrel, so the entire barrel and standing valve of the pump extend below the shoe. This arrangement is effective in low-fluid wells and wells that produce solids.

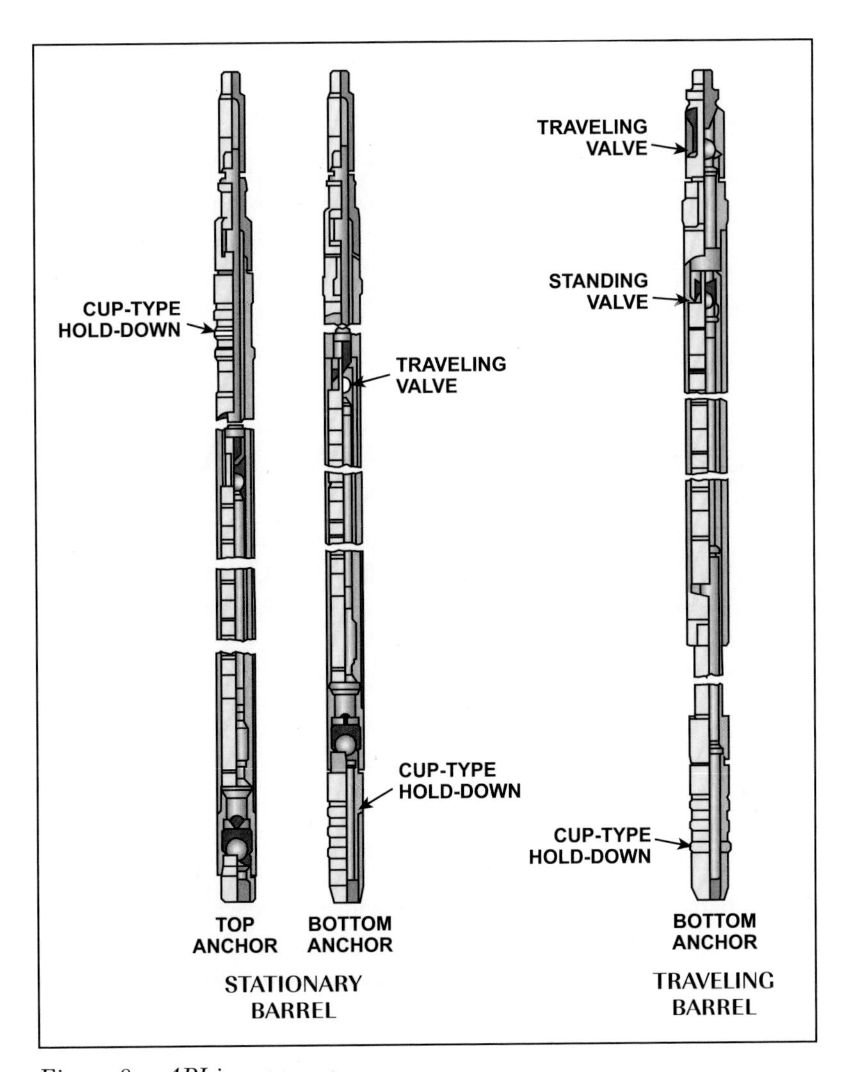

Figure 8. API insert pumps

- A stationary barrel bottom anchor pump has the hold-down on the bottom. The standing valve and entire pump are above the hold-down inside the producing tubing. This arrangement is effective in deep wells, but sedimentation accumulation around the barrel can cause the pump to become stuck.
- A traveling barrel bottom anchor pump has the hold-down on the bottom of the hollow pull tube below the plunger. The entire pump is above the hold-down and remains inside the production tubing. This arrangement is not recommended for deep wells or in the presence of gas. In fluids with high particulate matter, sedimentation can accumulate around the barrel when the pump is shut down, thus potentially creating a stuck pump condition.

Pump Components

Plungers can be manufactured to help resist abrasive wear that can reduce the lifecycle of a lift system.

The pump plunger connects to the rod string and acts as a piston in the pump barrel, lifting the fluid column on the upstroke when the traveling valve is closed. On the downstroke, the traveling valve unseats to allow fluid to pass through the hollow core of the plunger so that the plunger can fall through the fluid to the bottom of the pump barrel for the next upstroke.

Plungers can be divided into two groups: metallic and soft-packed (fig. 9). Metallic plungers rely on fluid film to fill the diametrical clearance between the plunger and barrel—a gap that can be as narrow as 0.002 inches (0.051 millimetres). This creates a seal and provides sliding lubrication. *Slippage* should be approximately 5% of the produced fluid volume to have adequate lubrication. The durability of metallic plungers makes them preferred for most applications. For increased abrasion resistance, plungers can be either chrome-plated or spray-metal-coated with a nickel-based alloy. Metallic plungers sometimes have circumferential grooves to trap debris from the fluid film to help prevent damage from abrasion.

Soft-packed plunger elements are nonmetallic and can be ring, split-ring, cup, or pressure-actuated types. Ring and split-ring sealing elements are assembled together onto a mandrel so that they fit snugly between the plunger and the inside diameter of the barrel. Cup plungers consist of a series of bowl-shaped, nonmetallic elements.

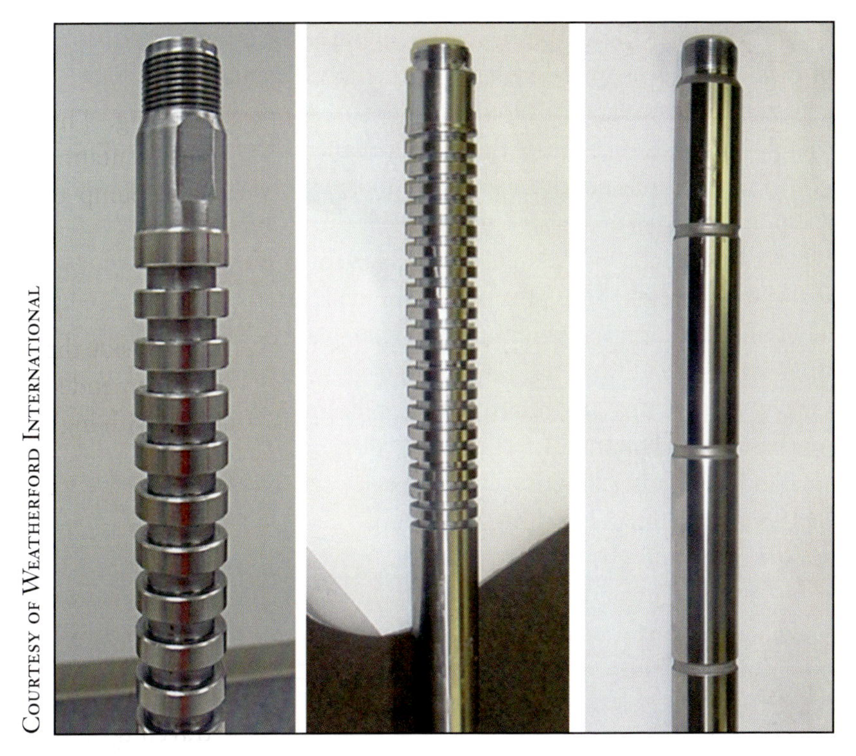

Courtesy of Weatherford International

Figure 9. Plungers

The elements face the direction of the pressure on the upstroke. As the cup plunger moves upward, the fluid pressure expands the lips of the cups against the barrel, forming a seal.

On the downstroke, the pressure on the cups is decreased, permitting the free downward movement of the plunger. Cup plungers are widely used for shallow to moderately deep pumping, usually not in excess of 5,000 feet (1,524 metres). The number, fit, and hardness of the cups are dependent upon the application depth, fluid, and other factors. Soft cups are used for very shallow pumping. Medium cups are designed for depths around 2,500 feet (762 metres); hard cups are required for deeper pumping (greater than 2,500 feet or 762 metres). Cup plungers can experience high wear when exposed to fluids containing sand.

The balls, seats, and cages are essential to pump longevity. They can be manufactured with special materials to resist the abrasive particles that are commonly found in fluids.

Pressure-actuated plungers use nonmetallic rings installed in individual grooves in the unit. As with cup elements, the pressure-actuated elements face the direction of the pressure on the upstroke so that the fluid pressure causes the sealing elements to seal more tightly.

Pump barrels are the largest and most expensive component in rod pumps. Barrel materials are typically steel or brass. They are most often nitrided, chrome-plated, or nickel carbide-plated for hardness and abrasion resistance. The barrels are made from precision tubing wherein the bores are honed to size after plating. In deep wells or wells where conditions are severe, a heavy-wall barrel can be used. Heavy-wall pumps are extremely rugged and will withstand high mechanical or hydraulic stresses. The thicker barrel walls result in smaller barrel bores, so the volume capacity of heavy-wall pumps will be less than comparable thin-wall pumps.

The standing valves and traveling valves are *ball-and-seat check valves*, which only permit flow in the up-hole direction. The balls and seats, along with the cages that contain them, are the most critical components in the pump with respect to pump life. Balls and seats are made from stainless steel or from special wear-resistant materials such as tungsten carbide and nickel carbide. Similarly, cages can be inserted with abrasion-resistant materials to extend life in abrasive fluids. Cages can also be made with oversized passages to reduce turbulence and flow resistance in wells with low fluid levels or high viscosity fluids, or those in which a high pump rate is required.

Double valves—two valves in tandem, one above the other—can improve pump performance in fluids with particulate matter such as sand or scale. If particles get between the ball and seat of one valve to prevent the ball from fully sealing, the second valve can close to hold the fluid column. Double valves increase the unswept volume within the pump, so they are not recommended in gaseous wells where gas interference or gas lock could occur.

In wells where calcium carbonate or similar deposits occur on the pump parts, the pump can be assembled as a stroke-through assembly. Extension nipples of a slightly larger internal diameter are added to each end of the working barrel so that the pump will stroke out of the bottom and the top end of the working barrel to keep it wiped clean.

Gas Anchors

Gas anchors, mud anchors, and gas separators are terms for subsurface devices that improve the efficiency of gravity separation in gaseous wells. They provide a chamber for gravity separation, discharge outlets for gas migration to the annulus, and outlets for liquid discharge to the pump intake. The chamber should have more volume than the pump displacement so that the fluids within have time to separate on the downstroke of the pump (fig 10).

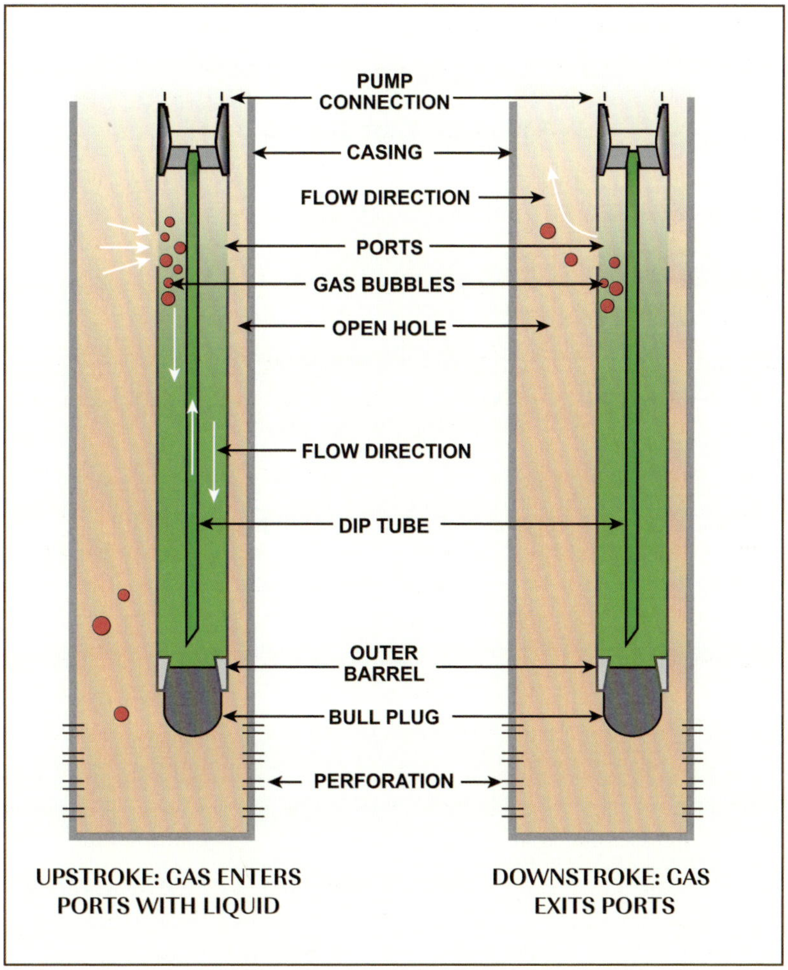

Figure 10. Gas separator process

Pump Nomenclature

API Specification 11AX defines nomenclature for subsurface pumps and fittings. A complete pump call-out records the following in order: the tubing size, pump bore, barrel length, extension length on top, and extension length on bottom (fig. 11).

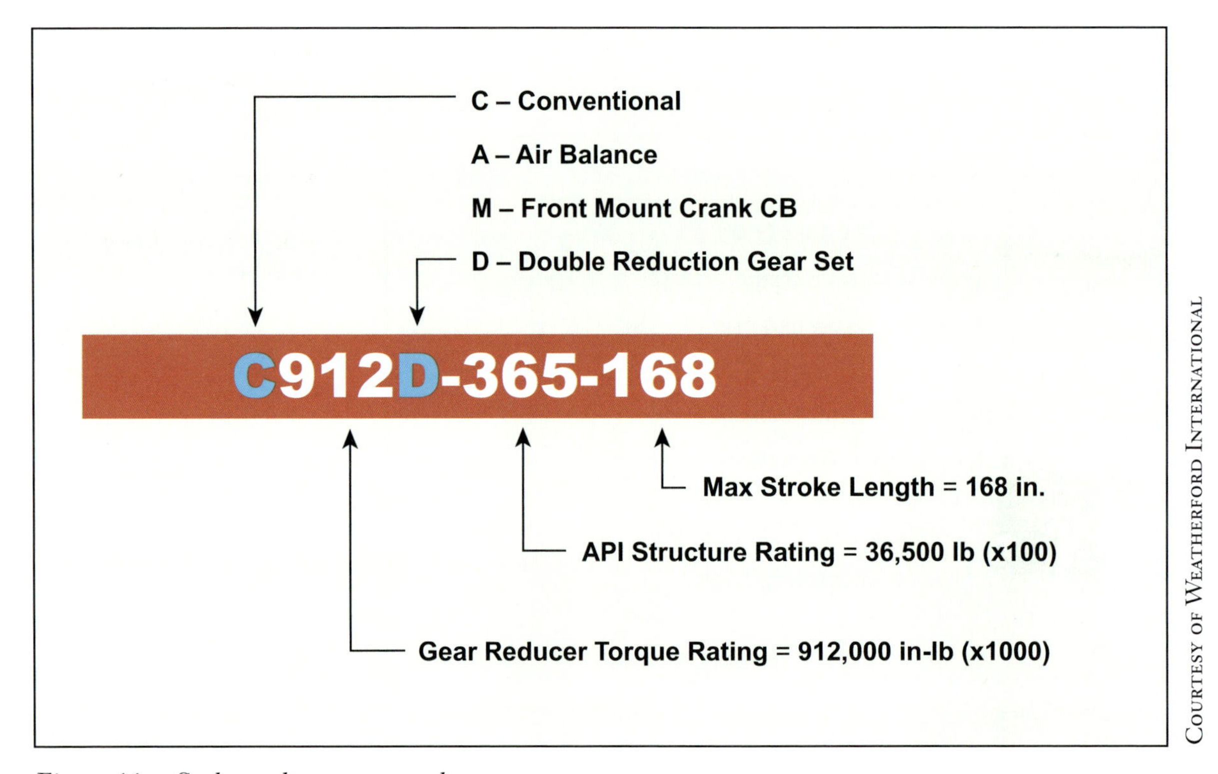

Figure 11. Sucker rod pump nomenclature

Surface Rod Pumping Units

Surface rod pumping units impart a *reciprocating motion* to the rod string, which then actuates the subsurface rod pumps. Pumping units are usually driven by electrical motors or by internal combustion engines using belts and sheaves (fig. 12). Typically, these mechanisms represent 70% of the capital cost associated with reciprocating rod systems and significantly impact operating efficiency.

Classification of surface rod pumping units:

- Conventional beam
- Front-mounted geometry crank counterbalance
- Phased crank counterbalance
- Beam balanced, long-stroke
- Low-profile
- Hydraulic

Courtesy of Weatherford International

Figure 12. Modern conventional pumping unit

Conventional Beam Units

Conventional beam units have a rear-mounted gearbox with a crank-type counterbalance to offset the weight of the rods and fluid (fig. 13). Conventional beam units are rugged, reliable, and considered to be the workhorse of the industry. The stroke length can have three or four adjustments, and rotation is bi-directional. The *counterbalance weights* are easy to adjust. The overall unit is easy to install, service, and repair. They retain a high resale value because of their long service life. (Reference: API RP 11E Rear-Mounted Geometry Class I, lever system with crank counterbalance.)

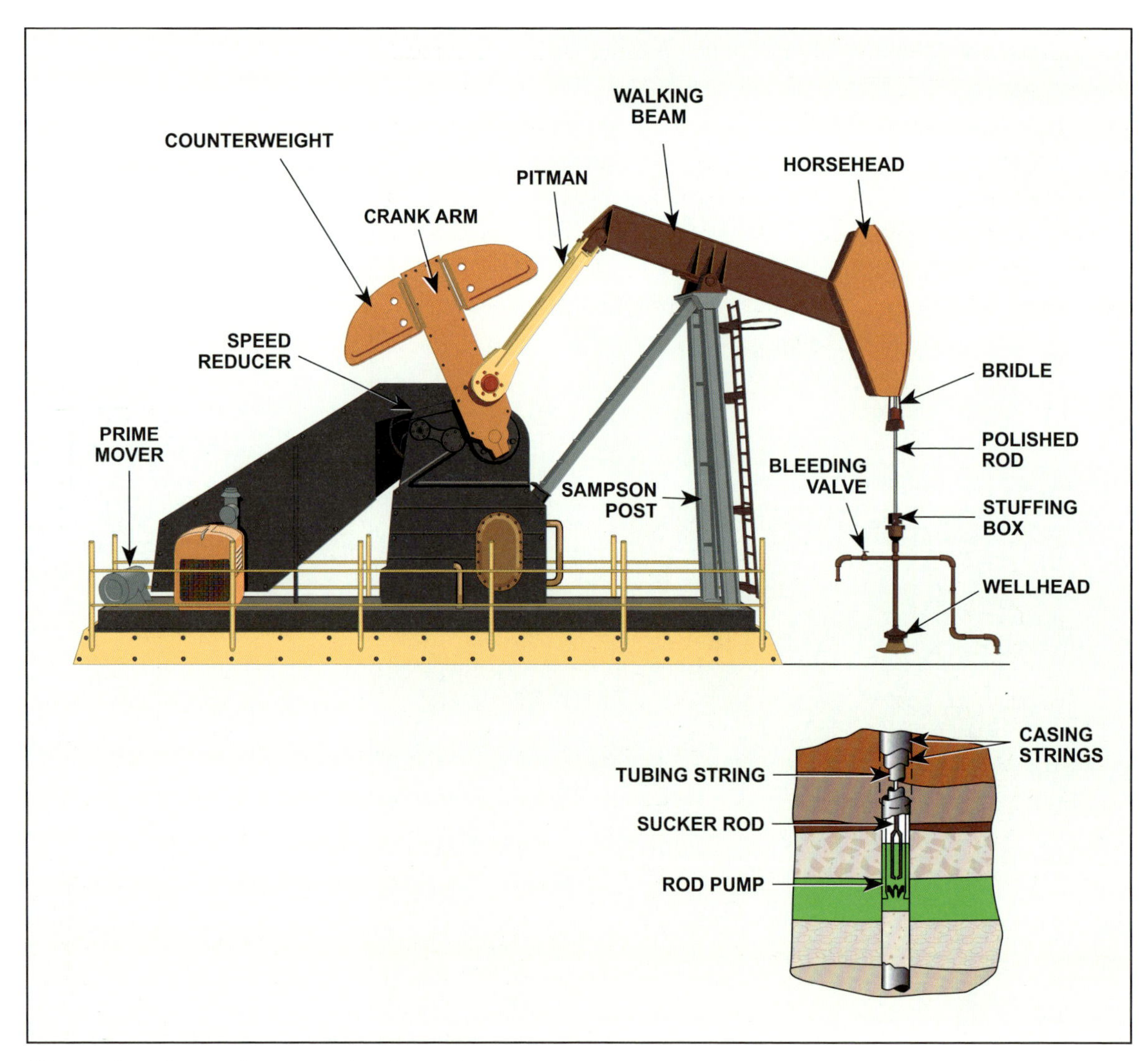

Figure 13. Parts of a conventional pumping unit

Front-Mounted Geometry Crank Counterbalance Units

Front-mounted geometry crank counterbalance units have the crank and *counterweights* positioned in front of the saddle bearing so that the pitman arms push up on the walking beam (fig. 14). The gearbox and crank are positioned to generate a slow upstroke and a fast downstroke. The resulting motion lowers peak gearbox torque, rod loads, and system horsepower requirements. The result is a more efficient pumping motion in some applications. In heavier fluids, deviated wells, and wells with flow restrictions, the faster downstroke might become a limiting factor for pump speed. In some cases, the unit must be slowed to prevent the rods from going into compression on the downstroke, which can lead to premature rod failures. The front-mounted geometry crank counterbalance units are typically more expensive than conventional units. Also, their unique layout provides limited space for the pumping unit to be driven by a gas engine. Specifically, the gearbox sits between the pumping unit frame and the wellhead with little room available for an engine. In addition, extra precaution is required for well service operations because the large counterweights in motion near wellheads can be hazardous. (Reference: API RP 11E Front-Mounted Geometry Class III, lever system with crank counterbalance.)

Courtesy of Lufkin Industries, Inc.

Figure 14. Front-mounted geometry crank counterbalance unit

Phased Crank Counterbalance Units

Phased crank counterbalance units have the layout advantages of a rear-mounted gearbox (as on conventional pumping units) while generating a slower upstroke and faster downstroke similar to front-mounted geometry crank counterbalance units (fig. 15). Compared to the front-mounted geometry counterbalance units, the difference in downstroke and upstroke speeds is less pronounced; therefore, the downstroke is less likely to put the rod string in compression. Also, the phased crank counterbalance units are easier to power with gas engines and are safer for well service operations due to the rear mounting of the gearbox, crank, and counterbalance. (Reference: API RP 11E Rear-Mounted Geometry Class I, lever system with phased crank counterbalance.)

COURTESY OF WEATHERFORD INTERNATIONAL

Figure 15. Phased crank counterbalance unit

Beam Balanced Units

Beam balanced units are the simplest and smallest rod pumping units (fig. 16). The crank and pitman arms are located behind the saddle bearing as with conventional units, but the counterweight is loaded on the rear of the walking beam rather than on the crank arms. These units are effective on shallow and low-volume wells.

Long-Stroke Pumping Units

Long-stroke pumping units provide reciprocating linear motion of 30 feet (9 metres) or more per stroke by using cables, long-stroke cylinders, or chain drives (fig. 17). The advantages of long-stroke units are higher production capacity, high system efficiency, variable upstroke-downstroke speed combinations, reduced rod loads, smaller footprint, and fewer system cycles. However, long-stroke units are typically tall and visually obtrusive and therefore might not be acceptable in some locations.

Courtesy of Weatherford International

Figure 16. Beam balanced unit

Courtesy of Weatherford International

Figure 17. Long-stroke pumping unit

Low-Profile Pumping Units

Low-profile pumping units are designed for use under agricultural watering systems (fig. 18). They are typically more expensive than conventional units, but they often have robust features including nitrogen over hydraulic counterbalance and variable speed control.

Courtesy of Weatherford International

Figure 18. Low-profile pumping unit

Hydraulic Pumping Units

Hydraulic pumping units are becoming increasingly popular because of their smaller footprint compared to other types of pumping units (fig. 19). Some units mount directly onto the wellhead without needing a concrete pad. The smaller footprint makes these units applicable in permafrost and other difficult locations. Lightweight units are easy to deploy and relocate, so they are used to test initial production or to temporarily produce wells that are expected to have steep production decline curves. The hydraulic unit controls allow easy, independent fine-tuning of variable upstroke and downstroke rates. Hydraulic units cost more than conventional units and require more maintenance.

Courtesy of Weatherford International

Figure 19. Hydraulic pumping unit

Rod String

Most operational failures that occur in reciprocating rod systems are related to issues with rods.

The rod string transfers the energy and motion of the surface pumping unit to the subsurface pump. The rods must operate under cyclic load conditions in erosive and corrosive environments, so it is not surprising that the majority of operational failures in reciprocating rod systems are related to rod issues. The design of the rod string and the proper makeup of the rods and couplings are the most critical success factors for a rod pumping system.

The elasticity of the rod string must be considered during the design of the reciprocating rod system. Incremental rod stretch on the upstroke can decrease the subsurface pump stroke relative to the surface pumping unit stroke. At some pumping speeds, the elasticity of the rod string can actually increase the subsurface pump stroke relative to the surface stroke because of system harmonics.

Because of the cyclic forces and stresses, the rod string must be designed and sized to resist fatigue failure. Metal flaws and minor physical damage to rods create stress concentrations, which can contribute to premature rod failures under cyclic stresses. Most of the nicks, dents, bends, and notches on rods are caused by improper handling. Bending of rods followed by straightening creates *tensile stresses* that can initiate microscopic cracks. Such damage is permanent, and early failure is almost certain.

The rods can be overstrained beyond their elastic limit when pulling on a stuck pump. One method of checking on strain is to measure the rod to see if it has been permanently stretched due to the pull. Damaged rods cannot be repaired and must be discarded.

Rod connections should be made correctly using the API circumferential displacement method. As the rod loads increase, additional care and precision is required on the makeup of the sucker rod connections.

API Sucker Rods

Sucker rods are available in many variations that meet API standards. Non-API sucker rods, tailored for special applications, are also available.

API sucker rods are defined in API Specification 11B (1998). They are available in standard diameters of ⅞-inch to 1⅛-inch (22-millimetre to 29-millimetre) increments of ⅛-inch (3.18-millimetre) and in standard lengths of 25 feet (7.62 metres), although 30-foot-long (9.14-metre-long) sucker rods have become the norm in some markets (fig. 20–22). They are manufactured from tightly controlled metallurgies to achieve a favorable balance of strength, ductility and impact properties, and fatigue resistance. Sucker rods are shot blasted after undergoing forging and heat treatment to remove scale and to improve the skin strength of the material. After cleaning and inspection, the ends are formed on automatic machines and threaded by a cold rolling process. This method gives the threads improved fatigue resistance and leaves the flanks of the threads smooth and free from tears and imperfections. The mating couplings also have rolled internal

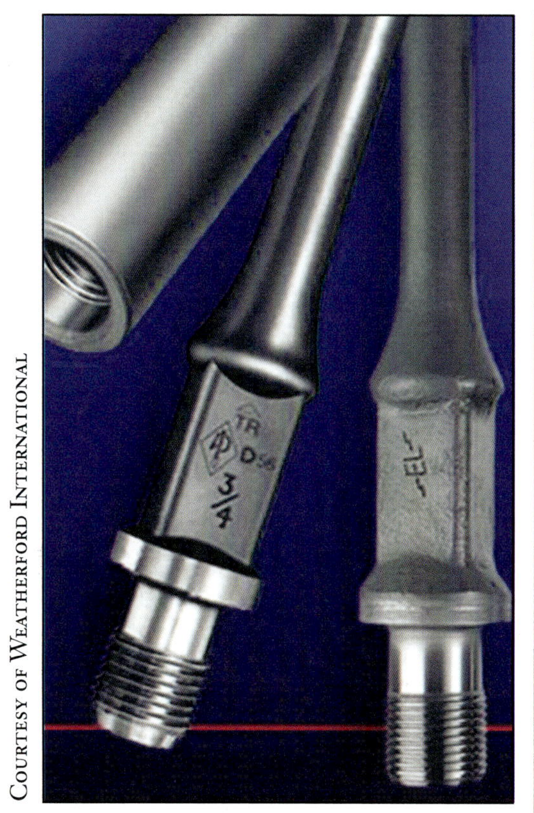

Courtesy of Weatherford International

Figure 20. Sucker rods

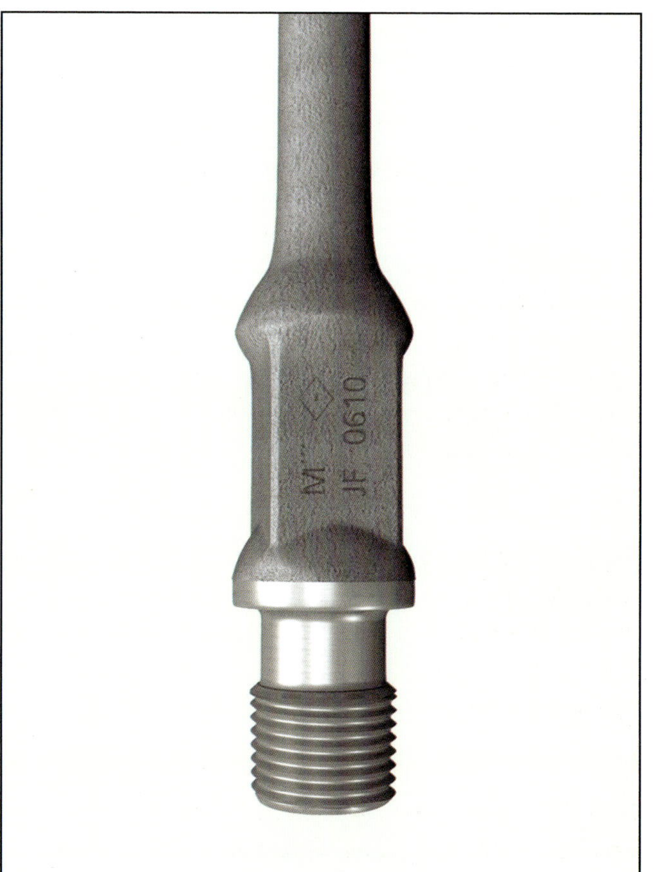

Courtesy of Tenaris

Figure 21. Sucker rod with an API pin

threads. API sucker rod grades, particularly in regard to *tensile strength*, generally have properties as shown in table 2. Although API defines minimum material properties, not all API rods are equal. Rods from manufacturers and different models from the same manufacturer vary in their yield strength, tensile strength, hardness on the Brinell scale, the heat treatment used to harden the metal, and other properties.

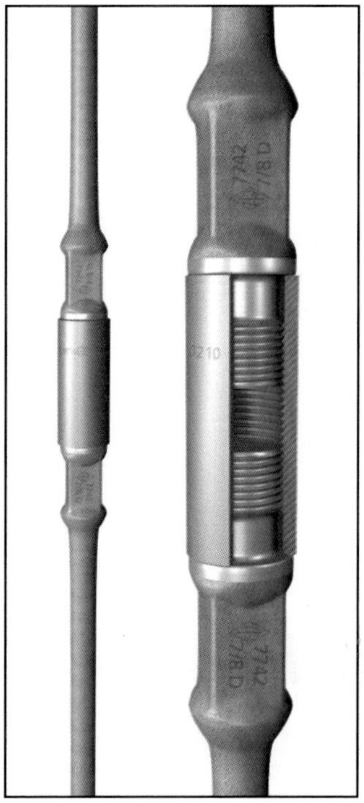

Courtesy of Tenaris

Figure 22. API sucker rods

Table 2
MECHANICAL STRENGTH PROPERTIES OF STEEL RODS

API Grade	Minimum Yield 0.2% Offset psi (Mpa)	Minimum Tensile psi (Mpa)	Maximum Tensile psi (Mpa)
K	60,000 (414)	90,000 (620)	115,000 (793)
C	60,000 (414)	90,000 (620)	115,000 (793)
D	85,000 (586)	115,000 (793)	140,000 (965)

Non-API Sucker Rods

Non-API sucker rods are available for extra-heavy loads, extra-deep wells, and corrosive environments. They are available in special materials and in diameters from ⅝-inch (16-millimetre) to 1½-inch (38-millimetre) to extend the rod performance range beyond what API rods can achieve.

Tapered Rod Strings

Tapered rod strings have more than one rod size (diameter) in the same rod string. Tapered rod strings are used to minimize rod weight in deeper wells. The tapered string weight is minimized by choosing the taper section lengths to maintain the same stress level on the upper-most rod in each taper section.

Non-API Connections

Non-API connections have been developed to extend the performance envelope of the rod string. Premium thread connections provide increased fatigue resistance and higher load capacity of rod strings. These high-performance connections require precision pin and coupling profiles, so they are substantially more expensive than API rod systems.

Fiberglass Rods

Fiberglass rods are lightweight and reduce the load on surface equipment. Fiberglass rods are available in standard lengths with steel pin end connections and steel couplings similar to conventional sucker rods. Fiberglass rods are suitable for corrosive environments, but in a typical design, half of the string is made of conventional steel sucker rods. The increased elasticity of fiberglass must be considered in the design process. Systems that use fiberglass rods can be designed to give the pump a longer stroke than the surface unit stroke. Compared to conventional sucker rods, fiberglass rods are more expensive, more susceptible to damage, and much more difficult to retrieve from the well if they break.

Hollow Rods

Hollow rods enable chemical injection through the rod string (fig. 23). The rods have non-upset internal thread connections and are joined together with external thread couplings. The flush connections minimize turbulence, reduce flow losses, and are less susceptible to overtorque than conventional sucker rods. The cost of hollow rods is significantly higher than conventional sucker rods.

Continuous Rod

Continuous rods eliminate the majority of box and pin couplings—the source of most rod issues.

Continuous rod consists of spooled rod injected into the well with only two rod connections: one at the subsurface pump and one at the *polished rod* on the surface (fig. 24). By doing away with most of the rod connections where sucker rods often fail, coupling failure can be almost completely eliminated. Continuous rod also reduces tubing wear by spreading the side forces along the continuous low-pressure contact with the tubing wall, in contrast to the concentrated higher-pressure contact that occurs with sucker rod couplings.

Elimination of couplings removes the coupling-piston effect in reciprocating rod systems and increases the flow area within the production tubing. Continuous rod strings are lighter than sucker rod strings, so they reduce the axial load at the surface. Continuous rods are particularly effective for deep, deviated, and slim-hole wells and for viscous fluid applications. The advantages of continuous rod must be evaluated in terms of the required capital investment and potential use of the equipment, because special equipment is required for deploying and retrieving this type of rod.

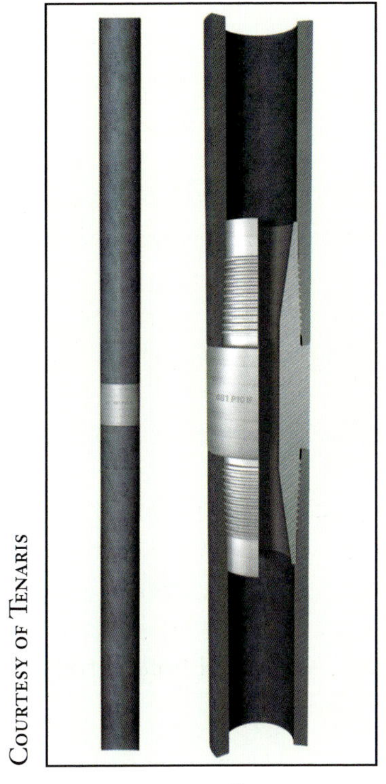

Figure 23. Hollow sucker rod

Figure 24. Continuous rod installation

Rod Guides and Poly-Lined Tubing

Rod guides are used to reduce friction and prevent wear between the sucker rod string and the production tubing in non-straight and deviated wells (fig. 25). Guides can be molded onto the sucker rods or installed in the field. Rod guides are not required on continuous rod strings. Alternatively, poly-lined tubing can be an effective way to reduce rod-on-tubing wear.

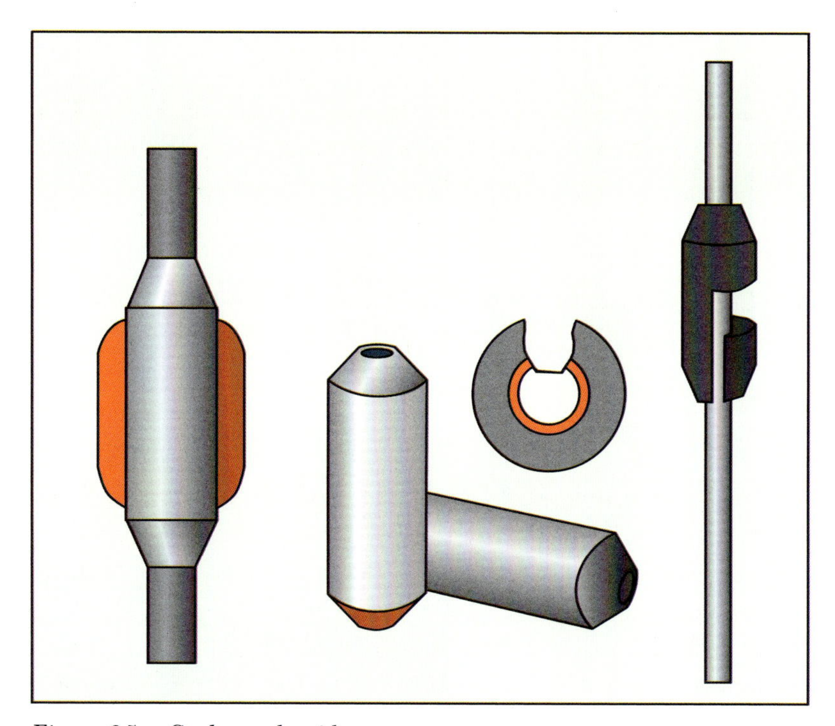

Figure 25. Sucker rod guides

Rod Handling

While running or pulling rods, proper handling is extremely critical. Pins and couplings should be kept clean at all times. The rods should be handled and stored to prevent bending (fig. 26). They should be stored off the ground with supports between the layers of rods at 6-foot (2-metre) intervals. To avoid bending during transit, the rods should be similarly supported by wooden spacers and secured with tie-down straps over the spacers. Equipment should never be placed on the rods. Operators should never allow a chain or hammer to contact a sucker rod. Rods should not be stacked in the derrick.

Courtesy of Weatherford International

Figure 26. Stacking rods on a rack off the ground prevents bending.

Rod Make Up

Threaded connections should be tightened sufficiently to preload the threads with higher stress than operating forces would cause. Preloading helps improve fatigue resistance by reducing alternating stresses. In the case of sucker rods, the preload places the rod pin in tension and brings the coupling into compressive contact with the rod pin shoulder. As long as this contact is maintained, the pumping load is carried by the whole connection. If the faces of the coupling and shoulder separate, the entire pumping load is transferred to the rod pin, usually resulting in fatigue failure. Sucker rod joint pin failures usually occur at the end of the thread runout where the thread blends into the pin stress relief.

Proper preloading is attained by controlling the make-up circumferential displacement. For ¾-inch (19-millimetre) rods and larger, power tongs can be used in conjunction with circumferential displacement cards. For ⅝-inch (16-millimetre) rods, hand wrenches are typically used. (Reference: API publication RP 11BR, Recommended Practice for the Care and Handling of Sucker Rods.)

System Design

General rule: Start with the longest stroke, the slowest speed, and the smallest plunger to produce required volumes.

Rod pumping system design is an iterative process whereby the system is initially sized to achieve the target production levels. The design is then refined to achieve an acceptable balance between competing criteria. The first step in the design process is to choose a subsurface pump size and stroke rate to deliver the target production rate. If there is a pumping unit on the well, then the stroke is defined by the pumping unit stroke. Otherwise, the general rule is to start with the longest stroke available, the slowest speed, and the smallest plunger size that will produce the required volumes. This combination will result in the lowest forces on the system components.

As a starting point, use a 24-inch (610-millimetre) stroke for a pump located at 2,000 feet (610 metres) true vertical depth. Then, add 10 inches (254 millimetres) of stroke for each 1,000 feet (305 metres) below that depth. An example would be a 64-inch (1,623-millimetre) stroke length for a pump located at 6,000 feet (1,829 metres).

Stroke speed must be chosen to avoid the *synchronous speed* of the rod system. A sucker rod string vibrates like a free-hanging coil spring. When the string is vibrated, the motion will travel up and down the string in waves with a specific velocity and frequency depending on its length. When the reflected waves meet other waves in such a way as to reinforce each other, the motion is said to be synchronous, and the resulting vibrations are amplified. The result can be large harmonic forces and stresses in the rods and on the equipment, which can lead to premature wear or failure of system components. If the synchronous speed is within the operating range of the pumping system, it can be identified by observing increases in peak rod loads as the pumping speed is adjusted.

The pump setting depth must be determined in order to define the rod weight, the pressure on the pump plunger, and therefore the forces in the rod string. When gas is present in the system, the pump should be set below the perforations by at least 50 feet (15 metres) to allow gravity gas separation prior to liquids entering the pump. For this reason, the artificial-lift system design should be considered during the drilling and well construction phases of field development. Gas anchors can be used to further help prevent gas from entering the pump. Alternatively, pumps in gaseous fluids can be located 200 feet (61 metres) above the pump to allow gas to break out of the fluid before entering the pump, although some gas would be expected to enter the pump. Locating the pump above the perforations will decrease the achievable pressure drawdown at the perforations.

The system power is determined by the depth from which the well fluids must be lifted, the volumetric rate at which they will be produced, the weight of those fluids plus adjustments for efficiency losses due to flow resistance, flow-line back-pressure, and pumping system efficiency factors. The fluid depth should be the total distance from the pump to the surface. The weight of the fluid is determined from the fluid specific gravity of those fluids. The total volumetric rate must consider all well fluids, including oil, gas in solution, condensate, and water.

Other necessary information includes:

- Whether the pump is set on a seating nipple, seating shoe, or pump anchor
- Fluid level in the well while pumping
- Static level if a new well
- Tubing size
- Whether the tubing is anchored or not
- Anticipated flow-line back-pressure

A trial rod string selection will be required to calculate forces in the system as part of refining the design. When selecting rod string, a common practice is to start with straight string (rods that are the same size) for pump depths to 4,000 feet (1,219 metres), a two-way taper for pump depths to 6,000 feet (1,829 metres), and a three-way taper for deeper pump seeting depths. Charts that show the component percentage of length for each size of rods used to make up a tapered string of sucker rods are available.

When selecting rod string for depths to 4,000 feet, a common practice is to start with a straight string—that is, rods that are the same size.

The design is then checked and adjusted using commercially available software based on wave equations and the Modified Goodman Diagram to determine stresses and projected rod life. These software programs use service factors to generate results that correlate to real operating conditions, so an understanding of the well conditions and the corresponding implications of the assumed service factors is required. For example, in a corrosive environment, it is sometimes better to operate at a higher stress level than to go to the next rod grade since stronger rods can be more susceptible to corrosion. It should be obvious that field experience is critical in selection of service factors and interpretation of the analysis results.

The design is based on the assumptions used in the analysis. The fluids and well conditions actually encountered are often different from what was assumed. Therefore, one must assume an appropriate amount of conservatism in the design process to account for any surprises in real well conditions compared to expected conditions.

Summary

Reciprocating rod lift, also called sucker rod pumping, is the most widely used form of artificial lift and is among the most efficient. Reciprocating rod systems are simple, efficient, rugged, and require little attention. They can effectively lift a wide range of liquids—from 5 to 5,000 barrels per day. Special configurations might be required for applications involving sand, gaseous fluids, and deviated well geometries.

The subsurface sucker rod pump consists of a plunger and two valves, a traveling valve within the plunger, and a standing valve at the base of the pump. The reciprocating motion of the plunger moves the fluid column toward the surface on the upstroke. On the downstroke, the traveling valve opens, letting fluid fill the chamber above the plunger. There are two types of rod pumps: tubing pumps and insert pumps. Surface rod pumping units are classified as conventional beam units, front-mounted geometry crank counterbalance units, phased crank counterbalance units, beam balanced units, long-stroke pumping units, low-profile pumping units, and hydraulic pumping units. The rod string is exposed to demanding conditions and stresses and is the component most likely to fail over time. System design is the process of matching a lift system to achieve target production levels, and then checking and adjusting the system as needed.

Electric Submersible Pumps

In this chapter:

- Typical applications of electric submersible pumps
- Operating principles for high volumes of fluids
- Key system components and how they function
- Basic ESP system design factors to consider

In 1916, Armias Arutunoff developed the first cylindrical multistage electric submersible pump (ESP) for dewatering mines and ships. He formed the Russian Electrical Dynamo of Arutunoff Company (REDA) and applied the technology to oilwells, first in Russia and then in Germany. Mr. Arutunoff immigrated to the United States and installed the first ESP in the Western Hemisphere in a Phillips Petroleum well in Kansas in 1928. By 1938, approximately 2% of artificially lifted oil in the United States was lifted by REDA pumps.

Today, ESPs have become the preferred lift technology for many pumping applications, from shallow dewatering of mines to high-volume offshore oil production. High-temperature systems have been developed to allow ESPs to pump in applications traditionally serviced only by rod pumping systems. Special gas-handling features have made it possible to use ESPs in some gaseous well applications. As a result, more capital is spent on procurement of ESP systems today than all other lift technologies combined.

Typical Applications

ESPs share the high-volume and offshore artificial-lift markets with gas lift as neither technology requires rod strings that would interfere with the function of subsurface safety valves (fig. 27).

ESPs are well-suited to high-volume, offshore, deviated, and horizontal wells.

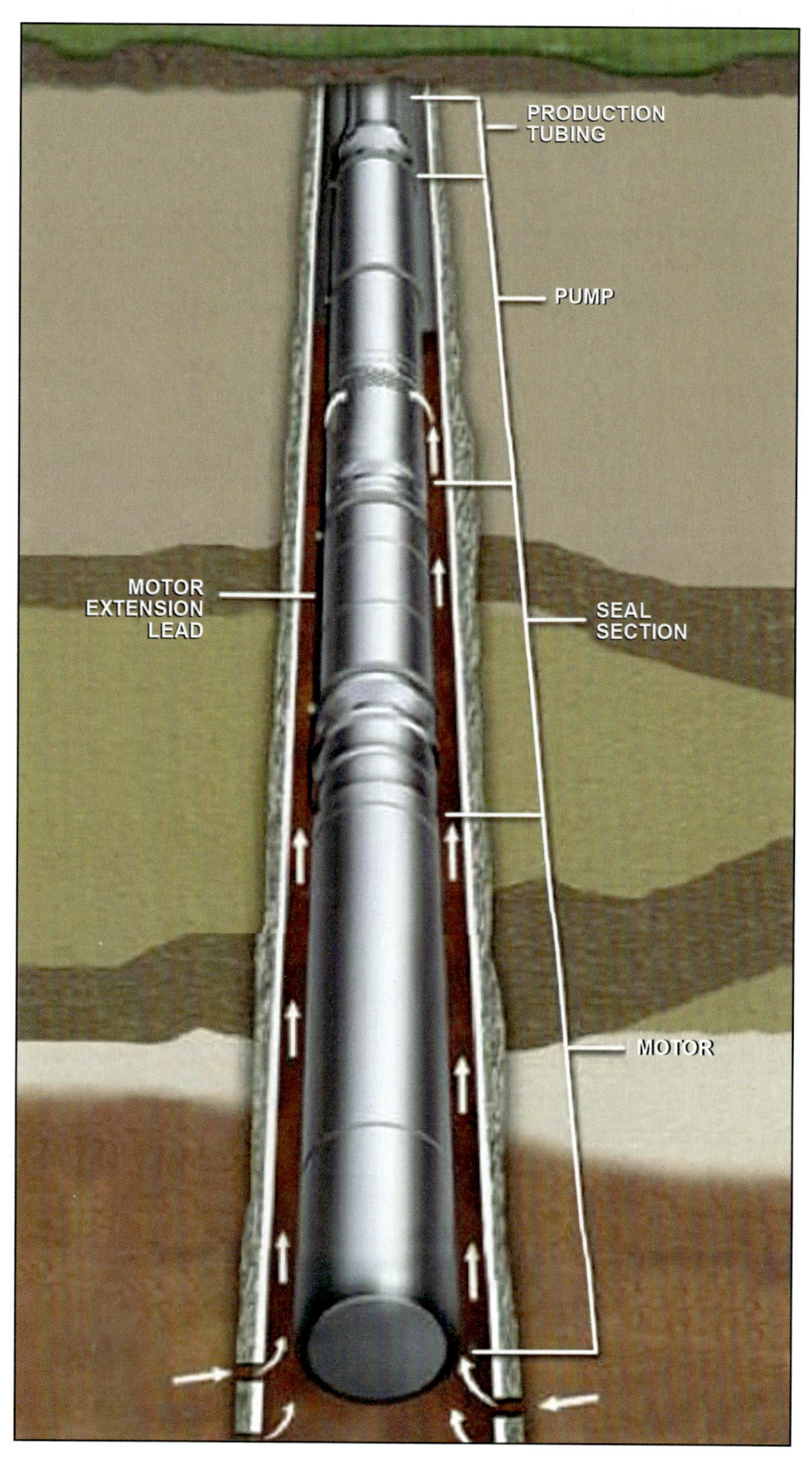

Figure 27. Electric submersible pump

ESPs are preferred over gas lift where required drawdown exceeds that which can be produced by gas lift, especially at higher water cuts and where electrical power is more readily available than compression. ESPs are also well-suited to deviated and horizontal wells. High-temperature ESP systems have been developed that are capable of withstanding internal motor temperatures greater than 500°F (260°C). The actual working temperature (BHT) limit for these systems would be 40°F to 60°F (4°C to 16°C) lower to allow for motor internal heating.

ESPs are ideal when:

- Electrical energy is inexpensive and readily available
- Flow rates do not fluctuate significantly
- Fluids are neither gaseous nor laden with particulate matter

The complexity and associated costs of ESP systems make them less competitive with other lift technologies in simple applications below 5,000 barrels per day, and they are usually uncompetitive below 1,000 barrels per day, depending on the well depth. The exceptions are simplified and economical ESP systems that are specifically developed for low-horsepower and shallow dewatering applications such as coal seam gas wells.

ESPs require high speed rotation in order to efficiently generate pressure. Therefore, they have a limited effective speed range or pump turndown ratio (fig. 28). Within that range, variable speed drives can be used to adjust ESP operating speeds to fit changing well conditions.

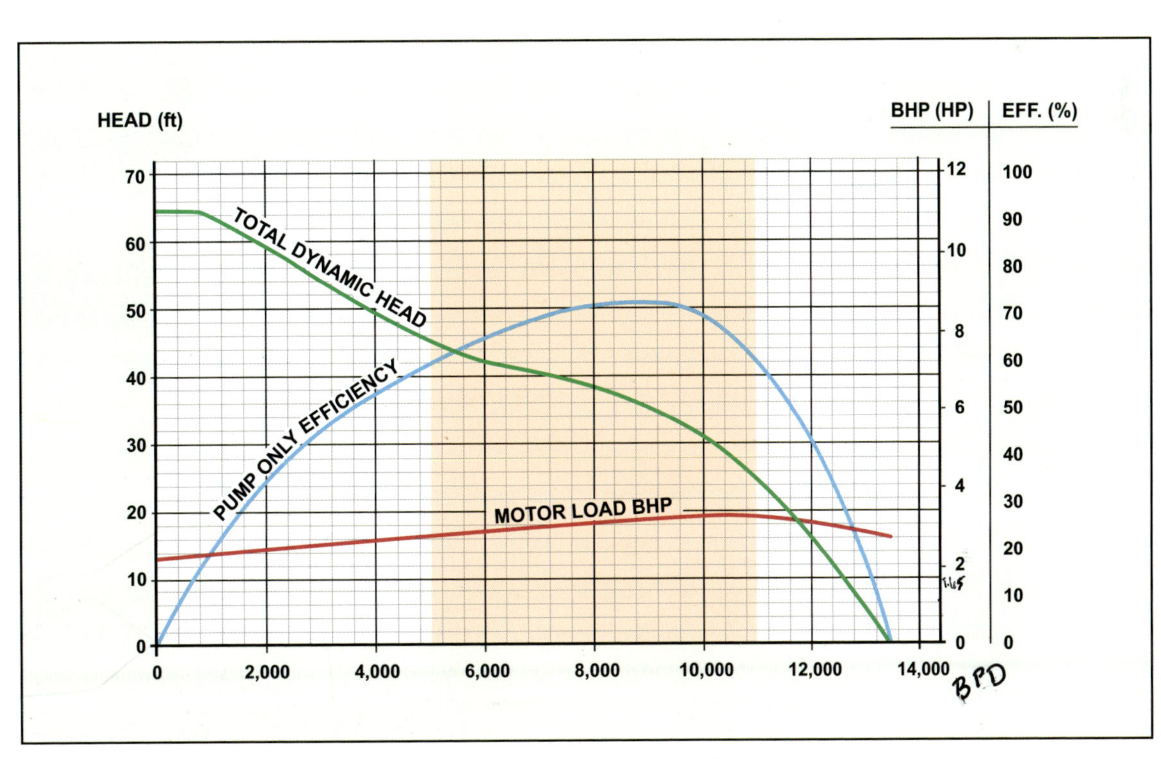

Figure 28. Graph of efficiency versus speed for electric submersible pumps

Courtesy of Weatherford International

Gas in ESPs reduces pump volumetric efficiency because compression of the gas decreases the fluid volume discharged from the pump. Gas can also accumulate within the stages to gas lock the pump so that no fluid is discharged. Gas separation devices can remove gas from the fluids entering the pump stages in applications that do not have high volumes of gas. Other ESP gas-handling devices are designed to create a gas emulsion or to compress the gas back into solution so that the well fluids are easier to pump.

ESPs are less efficient with higher-viscosity fluids (above 500 *centipoise*). Viscosity reducers, such as *diluents*, and thermal technologies can be used to lower viscosity of heavy oils. Particulate matter will limit the service life of ESPs. The torturous flow path within ESPs tends to cause larger particles to impinge against the *diffusers*, causing abrasive wear. Smaller particles can cause grinding wear in the bearings and shafts.

In general, ESPs can produce any volume of medium to light fluids wherever electrical energy is inexpensive and readily available, where flow rates do not fluctuate significantly, and with fluids that are neither overly gaseous nor laden with particulate matter. Most hydrocarbons are lifted to the surface using ESPs, and special high-temperature systems can be used in many thermal operations.

Operating Principles

ESPs are multistage centrifugal pumps that use centrifugal force to pressurize fluids (fig. 29). Within each pump stage, the vanes on a rotating *impeller* fling fluid outward due to centrifugal force. The fluid is then channeled through a diffuser to the intake of the next pump stage. When the fluid is pushed outward by centrifugal force, it leaves a low-pressure area at the center of the impeller, which draws in fluids from the preceding stage or the pump inlet. In this way, each stage draws in fluid from the preceding stage, boosts the fluid pressure, and channels the fluid to the next stage.

The subsurface pump assembly includes a(n):

- Electric motor to provide rotation and torque to the pump
- Seal chamber section, consisting of a barrier to keep well fluids out of the motor and a thrust bearing to support the force of the pump shaft
- Intake or gas separator
- Centrifugal pump

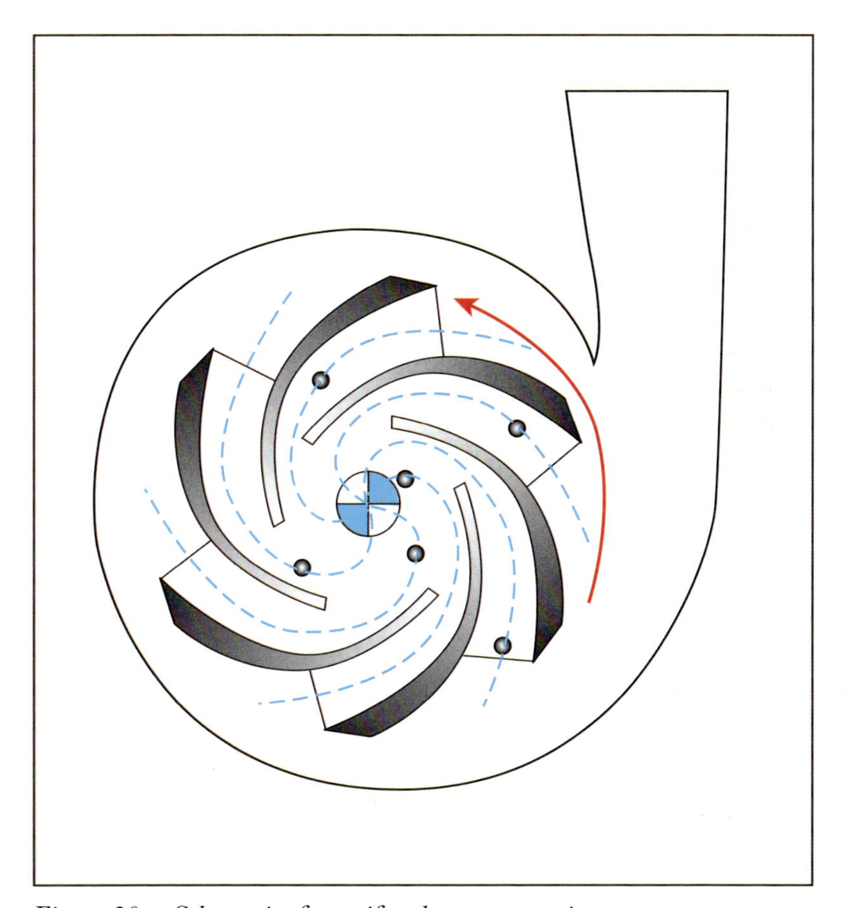

Figure 29. Schematic of centrifugal pump operation

Other potential components include chemical injection systems, subsurface gauges, and check valves. The pump assembly is attached to the bottom of the production tubing and electrically connected to the surface power equipment by means of a cable, which is typically strapped to the outside of the production tubing. The surface equipment includes a transformer, switchboard or variable speed controller, a wellhead adapter or special wellhead that accommodates cable penetration, and a vent box to release gases that migrate up the electric cable.

System Components

Deep applications require pumps with several hundred stages of centrifugal impellers and diffusers.

ESPs consist of several key subsurface components, including a:

- Pump assembly
- Motor
- Motor seal separator
- Intake or gas separator
- Power cable
- Standing valve

Pump Assembly

Courtesy of Weatherford International

Figure 30. Impellers

The pump consists of multiple stages of centrifugal impellers and diffusers that are assembled vertically (fig. 30). Rotation of the impellers develops pressure on the fluid due to centrifugal force. Each stage provides additional pressure differential. The sum of these pressures—its magnitude depending on the number of stages—furnishes the pressure needed to lift the fluid to the surface. Pumps with several hundred stages might be required for deep applications. The pump pressure capacity is also dependent upon the rotational speed of pumping. The volume produced by the pump is determined by the displacement of one pump stage, the speed of rotation of the pump, and the fluid leakage, or slippage, between the stages.

Pump stages can be either for radial flow or mixed flow (fig. 31). Radial flow stages are short and provide the most compact pump assembly, but they are less efficient and more susceptible to gas locking in the presence of high GLR fluids. Mixed flow stages are longer and less efficient than radial flow stages except in high GLR environments because they are less susceptible to gas locking. They are more effective at flushing gas from one stage to the next and for moving particulate matter through the pump. The longer stages cause mixed flow pumps to be longer and more expensive than radial stage pumps.

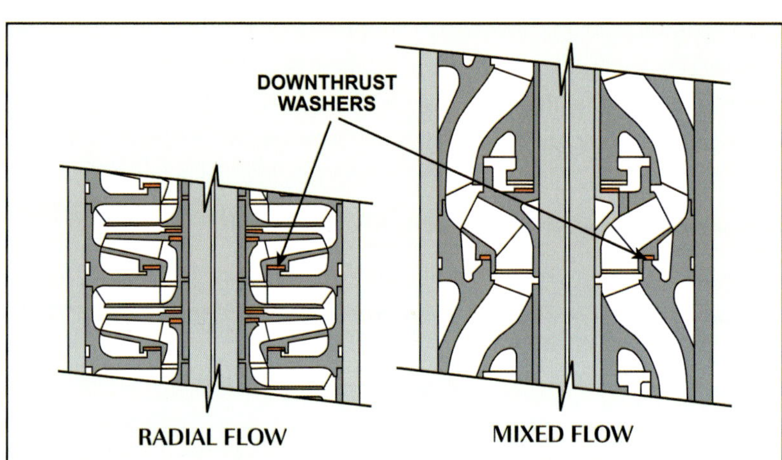

Courtesy of Weatherford International

Figure 31. Radial flow and mixed flow stages

The downward thrust on the pump created by the discharge pressure can be transferred through the shaft to the thrust bearing (compression pump), or the stages can be configured for each stage to transfer thrust to the diffuser and into the pump housing (floating pump) as shown in figure 31. Compression pumps eliminate the impeller-to-diffuser wear interfaces, so they are more tolerant of abrasives than are floating stage configurations.

Certain materials can be used to prevent wear caused by particulate matter. Still, particulate matter is likely to cause at least some damage to the pump over the course of an ESP's lifecycle.

Motor

The electric motor is typically an induction motor that operates between 1,200 and 5,400 *revolutions per minute* when controlled by a variable-speed drive. Lower and higher speeds can be achieved by using special motor configurations such as high-torque motors. Most ESP systems operate on 480 and 380 VAC 3 (Volts Alternating Current) phase power, but other power configurations can be accommodated. The motors are enclosed in steel housings filled with dielectric oil that serves as both a lubricant and cooling agent (fig. 32).

Permanent magnet motors have also been developed that improve electrical efficiency and provide a broader speed range. The improved efficiency results in proportionately less internal heat generation.

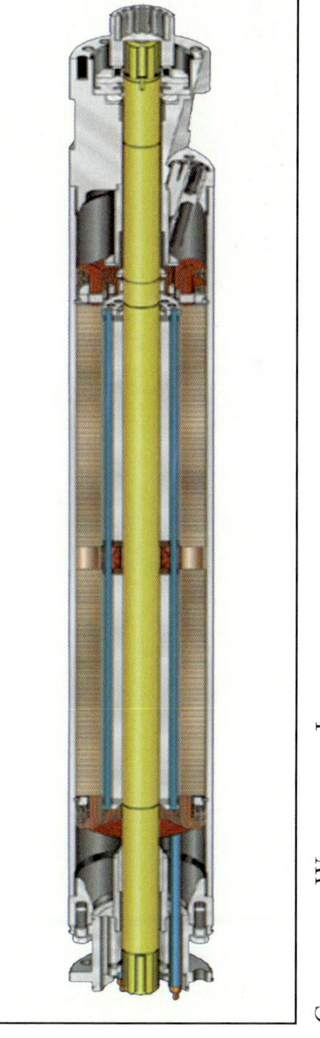

COURTESY OF WEATHERFORD INTERNATIONAL

Figure 32. ESP motor

Motor Seal Section

The seal section serves three purposes:

- To exclude the pumped fluid from the motor by equalizing fluid pressure within the motor with the pressure of submergence in the well
- To isolate the thrust of the pump from the motor bearings
- To permit the expansion and contraction of the motor and oil as it heats up during operation and cools off when the motor stops

Bag-type separators utilize elastic membranes to separate the motor oil from the well fluids. They are effective when varying temperature and flow conditions are expected, but they can be susceptible to chemical attack and degradation from prolonged exposure to elevated temperatures. Labyrinth-type seals are mechanical seals that separate the motor oil from the well fluids by means of a long and difficult fluid path (fig. 33). There is no elastomer involved, so labyrinth seals are less susceptible to chemical and temperature extremes. However, they can allow some comingling of miscible fluids as temperatures and pressures vary, especially with lighter produced hydrocarbons. Labyrinth seals should not be used at conditions close to the bubble point of the fluids because gas breakout of fluids in the labyrinth seal can displace the liquids in the labyrinth. Inclination also imposes a performance limit on the labyrinth seal. Some seal sections use both bag and labyrinth seal elements arranged so the labyrinth seal keeps the well fluids and production chemicals off of the bag seal.

Intake or Gas Separator

The intake can be integrated with the pump housing or it can be a separate bolt-on component. If either mechanical or static gas separation is needed, then a gas separator device is used as an intake. Most mechanical gas separators spin the fluid so that the heavier liquids move toward the outside of the housing, displacing the lighter gases toward the centerline of the housing (fig. 34). The gas is then vented outside of the separator while the liquid is channeled into the pump intake. Static gas separators can be used in deviated wellbores to take advantage of fluid gravity segregation and intake fluid from the low side of the tubing.

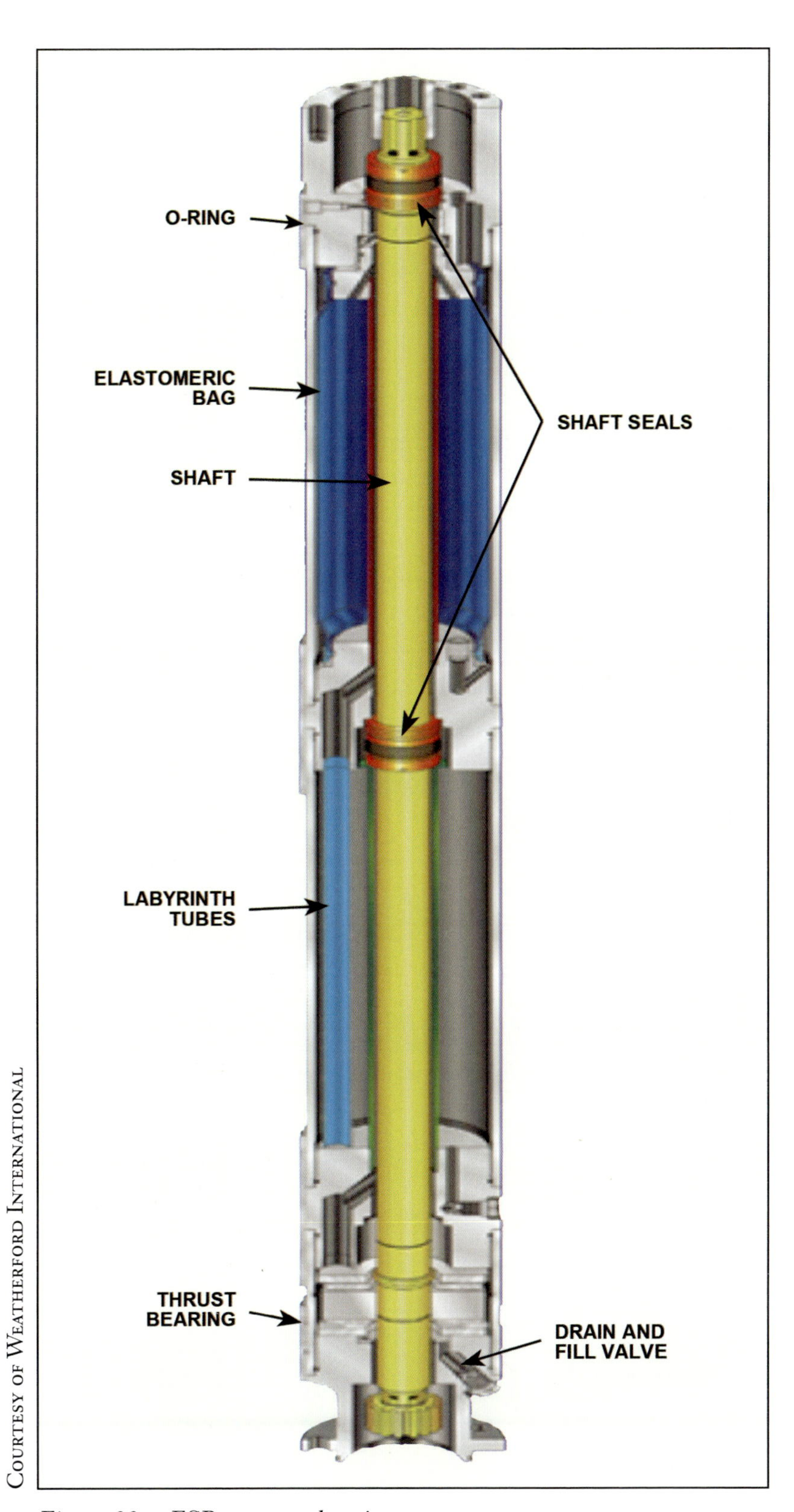

Figure 33. ESP motor seal section

COURTESY OF WEATHERFORD INTERNATIONAL

Figure 34. ESP gas separator

Some gas-handling devices agitate the fluid to break up larger gas bubbles into fine ones to help keep them flowing through the lower stages of the pump. This helps deter gas locking by preventing the bubbles from coalescing into pockets of gas within the stages. The size of the bubbles does not impact the percentage of gas in the fluid, so the compressibility of the gas and the resulting efficiency losses are unaffected by the smaller gas bubbles. Above the pump, the gas bubbles in the rising fluid column will expand as the hydrostatic head decreases. This makes the fluid column easier to lift. (A similar effect is achieved in gas lift.)

Power Cable

An ESP power cable should:

- Have a low profile
- Hold line losses to a minimum
- Resist pressurized well fluids
- Protect against mechanical damage
- Be flexible and spool easily

An electric submersible pump assembly must be connected to the surface equipment by a power cable. In deep wells, the cable can be the most expensive component in the ESP system. It must have a small diameter but must hold *line losses* to a minimum (fig. 35). The insulation must be resistant to pressurized oil and water and to gas impregnation. The cable must be protected against mechanical damage but has to be flexible for ease of spooling. Wire size depends upon the depth of the pump and the power needed. Improvements in insulating materials for the three individual conductors and the cable jacket permit use of cable at depths in excess of 10,000 feet (3,048 metres) and at high temperatures.

The power cable can be flat or round. Round cable is preferred where space permits because impedances between the wires remain balanced, and the cable can more easily roll to avoid damage at pinch points during installation and pulling of the system. Motor leads which connect the cable to the motor are usually flat due to space constraints. However, for some large ESPs and where space permits, round motor lead extensions are used. Cable is attached to the tubing by clamps spaced at 15- to 30-foot (4½- to 9-metre) intervals. A vented junction box at the surface is necessary to release any formation gases that migrate into the cable.

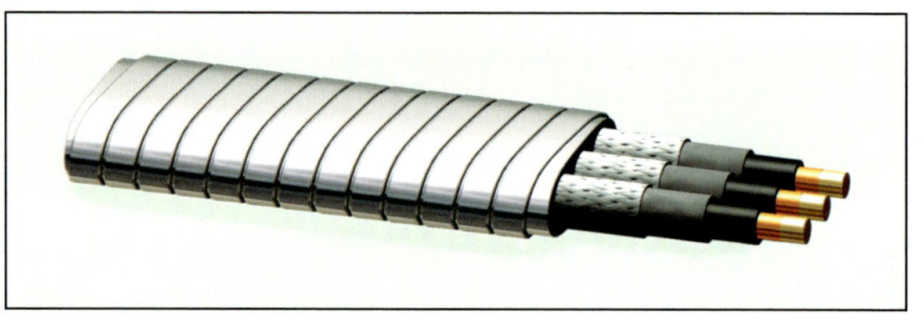

Courtesy of Borets

Figure 35. ESP power cable

Standing Valve

A subsurface check valve is sometimes used to hold the column fluid in the event of a system shutdown. This reduces the system start-up time, but it can create a high start-up load on the pump system.

ESPs are equipped with the following surface equipment:

- Electric controls
- Wellhead

Electric Controls

The most basic surface controls include a motor soft starter with circuit breakers or fused overload protection. Pump-off protection can be provided by sensors that detect low electrical current and then shut off power to the pump. Pressure, temperature, and vibration gauges can be used by the control to more accurately sense pump conditions. Signals from subsurface gauges are sent through the power cable. Recording ammeters, alarms, *SCADA* (*Supervisory Control and Data Acquisition*) systems, and other accessories can be included.

Variable speed drives (VSDs) allow the operating envelope to be adjusted to meet well inflow conditions rather than starting and stopping the pump to adjust production volumes. They also provide more options with respect to starting and shutdown sequences. The frequencies of induction motor VSDs can vary from 20 to 90 hertz, but 45 to 75 hertz is a common industry practice to match the practical operating envelope of the pumps.

Wellhead

Special ESP wellheads can be used to accomodate the power cable. These include a cable pack-off seal with electric feed through connections for the cable as shown in figure 36.

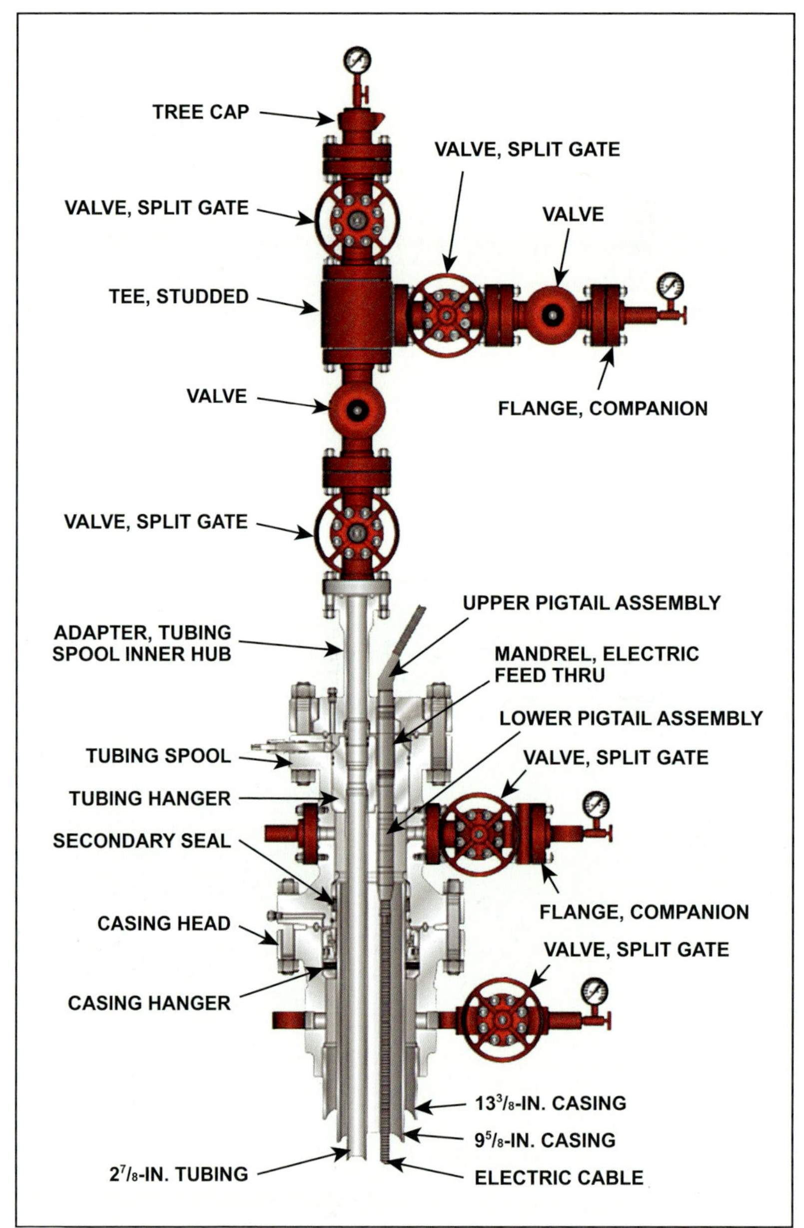

COURTESY OF WEATHERFORD INTERNATIONAL

Figure 36. ESP wellhead

System Design

When sized and configured accordingly, ESPs should overcome:
- Gas
- Particulate matter
- Back-pressure
- Flow losses

ESPs are sized by comparing the displacement of the pump stage size to the target production volume, and by choosing the number of stages to provide sufficient pressure to overcome:

- The hydrostatic head of the fluid in the tubing
- Back-pressure from the surface
- Flow losses

The production volumes can be easily adjusted within a narrow operating range outside of which the efficiency of the ESP drops off quickly. If changes in production volumes result in significant efficiency degradation, then the system might need to be resized.

ESP components can be configured to overcome specific challenges such as gas, particulate matter, and temperature as discussed in the preceding section. Locating the pump below the perforations will allow gravity to help keep gas out of the pump. Similarly, locating the pump above the perforations will allow particulate matter to fall without passing through the pump.

Shrouded Systems

A shroud can be attached to the pump intake to cover the seal section and the electric motor. The shroud channels the pump intake flow past the electric motor to help with motor cooling. This arrangement is particularly helpful in applications with low fluid velocities around the motor, a situation common to large-casing applications. Shrouded assemblies also allow the pump to be landed below the perforations to channel liquid past the motor for cooling while allowing formation gas to flow up the casing annulus prior to fluids entering the pump. There is also a configuration, called a recirculation system, which bleeds a small portion of the pumped fluid back down below the motor through a small diameter tube to help with motor cooling in well casings that are too small in diameter for a shrouded ESP system.

Bottom Intake Systems

The components can be assembled in reverse order with the electric motor on top, the pump on the bottom, and the seal section in between. Fluids are produced up the annulus, so a *packer* is required to segregate the pressurized annulus from the formation inflow. This arrangement provides good cooling for the motor but is not widely deployed except in special applications such as coil tubing deployed systems. Bottom intake systems are not recommended for gaseous wells because any gas must be produced through the pump. The casing is exposed to the well fluids and the full discharge pressure of the pump, so casing integrity must be considered.

Coiled-Tubing Deployed Systems

ESPs can be deployed and operated with *coiled tubing* without workover rigs. The subsurface assembly is suspended on coiled tubing in a bottom intake configuration with round power cable inside the coiled tubing. Fluids are produced up the annulus so the system has the same limitations as bottom intake configurations.

Summary

Electric submersible pumps (ESPs) are multistage centrifugal pumps. Within each stage, vanes on a rotating impeller move fluid outward and then upward to the next stage. The subsurface pump assembly includes an electric motor to provide rotation, a seal chamber section, an intake or gas separator, and the centrifugal pump. The power cable is another vital component. ESPs are the preferred lift technology for certain pumping applications, including pumping very high volumes of fluid. They are well-suited to highly deviated wells and offshore wells. However, most ESP systems are complex, and they are only efficient within a narrow envelope of operating parameters. ESPs are sized by comparing the displacement of the pump stage size to the target production volume and by choosing the sufficient number of pump stages. ESP components can be configured to overcome specific challenges such as gas, particulate matter, and temperature.

Progressing Cavity Pumps

In this chapter:

- Typical applications of progressing cavity pumps
- Operating principles and functionality
- System components that achieve high rates of efficiency
- Installation and operation factors to consider

In the late 1920s, René Moineau developed a new positive-displacement pump concept that was later commercialized into the first progressing cavity pumps. *PC pumps* were originally optimized for industrial applications to pump paper pulp, food, sewage, and other viscous fluids. As the oil industry began to produce more viscous and sandy fluids, PC pumps were found to have better performance than rod pumps and most other lift technologies used in the production of cold, heavy oil. Initially, oilfield PC pumps were industrial pumps installed vertically in the wellbore and driven by a rotating rod string turned by a surface drive system. Over time, special pump geometries and high-performance elastomers were developed to optimize production of various oil compositions including medium and lighter oils. Today, rod-driven PC pump systems typically have the highest operating efficiency of any lift technology (fig. 37).

Compared to other lift technologies, rod-driven pump systems are typically the most efficient.

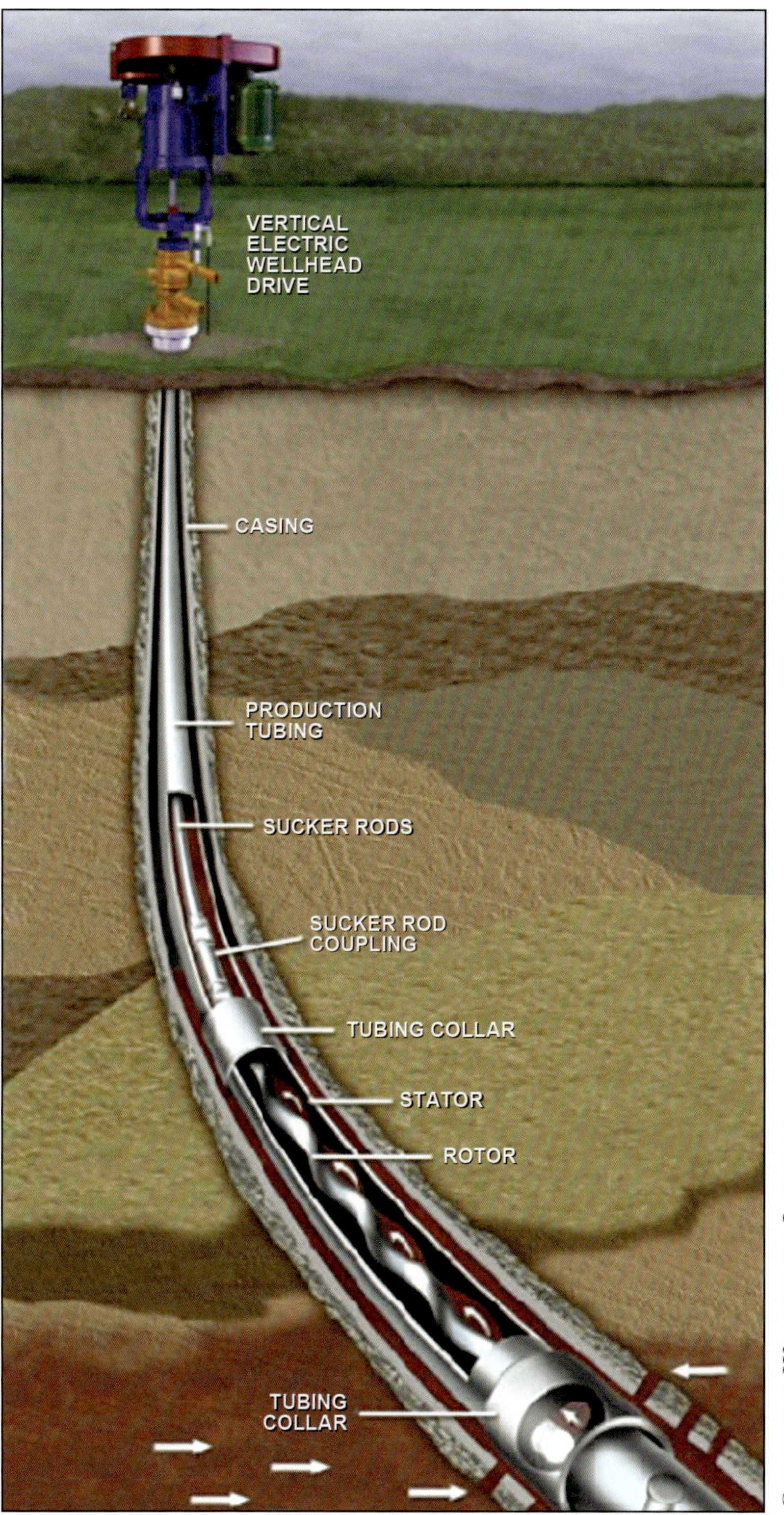

Figure 37. Progressing cavity pump

PC pumps consist of a corkscrew-shaped rotor placed inside a helical cavity stator. As the rotor turns, the configuration of rotor and stator creates sealed cavities that progress from the pump inlet at the bottom to the outlet at the top. PC pumps are positive-displacement pumps that hold pressure and support the fluid column at any operating speed (fig. 38). Conventional PC pumps are rotated by rod strings from a surface drivehead (fig. 39). PC pumps can also

Courtesy of Weatherford International

Figure 38. Cutaway view of a progressing cavity pump

Courtesy of Weatherford International

Figure 39. PC pump surface drivehead

be driven by subsurface, ESP-style electric motors for special applications. These bottom-drive PC pumps are sometimes referred to as *ESPCPs* (fig. 40).

COURTESY OF BAKER HUGHES INCORPORATED

Figure 40. Cutaway view of an electric submersible progressing cavity pump

Typical Applications

PCPs are ideal for pumping:
- Viscous fluids
- Abrasive fluids with particulate matter
- Many medium and light oils, depending on the chemistry of the fluids

Progressing cavity pumps are currently the industry-preferred, positive-displacement lift technology for pumping viscous fluids and abrasive fluids with particulate matter. They are extensively used for pumping cold, heavy oil and to dewater coal bed (coal seam) methane wells when coal fines—small particles of coal—are present. PC pumps have pumped 8° API extra-heavy oil with fluid viscosities as high as 40,000 centipoise. They have also intermittently pumped sand concentrations as high as 50%.

PC pumps are also effective in medium and light oils, depending on the chemistry of the fluids and their gas content. PC pumps are less susceptible to gas lock than other mechanical pump systems, as PC pumps have no valves or reciprocating components. However, special elastomers must be used to resist explosive decompression from prolonged exposure to certain gases such as CO_2. Likewise, special elastomers are required to resist attack from aromatic gases that can cause elastomer swelling or shrinkage.

Rod-driven PC pump systems typically have lower initial capital costs compared to other mechanical lift systems. Therefore, PC pump systems are used where operating efficiency and low initial capital costs are critical.

Currently, conventional PC pumps produce rates as high at 5,000 barrels of fluid per day. However, this is a function of well depth and casing size. High-rate systems typically require installation in large diameter casing and are landed at a relatively shallow depth of less than 3,500 feet (1,067 metres) true vertical depth. PC pumps driven by rod strings are depth-limited to around 8,600 feet (2,621 metres) due to the high head capacity requirement, which results in high rod stresses and extremely long pumps that make shipping and handling very difficult.

PC pumps are available today for 250°F to 300°F (121°C to 149°C) applications. Advances in pump technology are expected to move the temperature limit closer to 350°F (177°C) for elastomer-based pumps. Pumps with metal rotors and metal stators that do not include elastomers are less temperature-constrained but have other issues related to effectiveness with producing particulate matter.

Conventional rod-driven systems represent the vast majority of PC pump applications due to their simplicity, reliability, and efficiency. They utilize a rod string to transfer torque to a downhole pump. Because the rods interfere with subsurface safety valves, conventional PC pump systems are primarily used for land applications. PC pumps can be used offshore if fluid levels are sufficiently high to allow the pump to be landed above deep-set subsurface safety valves (SSSVs).

Unlike conventional PC pumps, ESPCPs can be used offshore and in deviated wells, since they have no rod strings. The ESPCPs can either be driven by conventional ESP motor and gearbox assemblies or by special high-torque electric motors. A connecting shaft is required between the drive system and the pump to accommodate the eccentric sideways translation of the PC pump rotor as it rotates. ESPCP systems are more complex than conventional PC pump systems, but they have an offsetting advantage of less flow resistance because there is no rod string in the production tubing.

Operating Principles

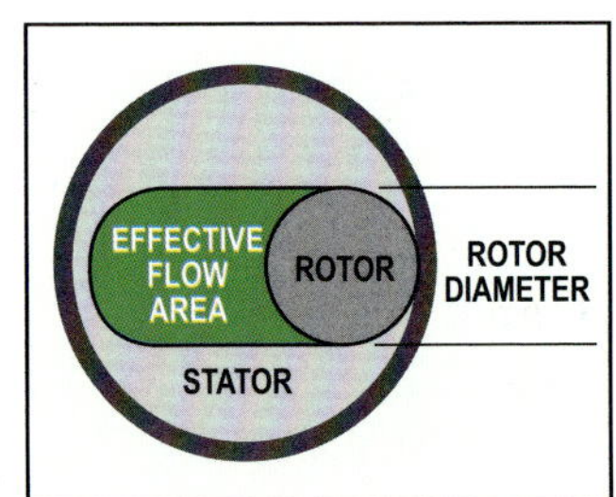

Courtesy of Weatherford International

Figure 41. Schematic of a PCP rotor, stator, and cavity

Courtesy of Weatherford International

Figure 42. PCP rotor in stator

The majority[3] of progressing cavity pumps consists of a circular cross-section helical (spiral-shaped) rotor that turns inside of a stator cavity that has a helical semielliptical cross-section (fig. 41). This stator cavity is described as two-lobe. In cross-section, it resembles a rectangle with rounded ends (fig. 42). The round, cross-section rotor is referred to as a single-lobe rotor, so the complete pump is referred to as a single-lobe or 1:2 pump configuration. The 1:2 stator cavity helix has twice the pitch of the rotor helix (fig. 43). The resulting assembly creates a series of sealed cavities between the rotor and stator that move upward, or progress, from the pump inlet to the pump outlet as the rotor turns. Although the cavities are not connected, they overlap longitudinally to create quasi-continuous flow through the pump (fig. 44). The resulting pumping motion is smooth without significant turbulence or fluid shearing.

The rotor motion consists of rotation about its axis plus translation sideways back and forth across the stator cavity. As a result, the rod string or drive rod connected to the PC pump must be free to translate sideways as well as rotate.

Multi-lobe 2:3 configurations consist of rotors with elliptical or other two lobe cross-section geometries inside of three lobe triangular cross-section stators. The 2:3 configurations have progressing sealed cavities similar to 1:2 configurations but have higher displacement and therefore move more fluid volume per revolution compared to 1:2 pumps of similar length. The higher elastomer flexing frequency of the 2:3 configurations can lead to reduced pump reliability (fig. 45).

A surface drivehead transfers energy to the subsurface rotor by means of a rod string. In this respect, PC systems are similar to reciprocating rod systems except the rods rotate rather than reciprocate. The surface drivehead can be powered by electric motors or

[3] Over 95% of PC pump applications use 1:2 configuration pumps as of 2010.

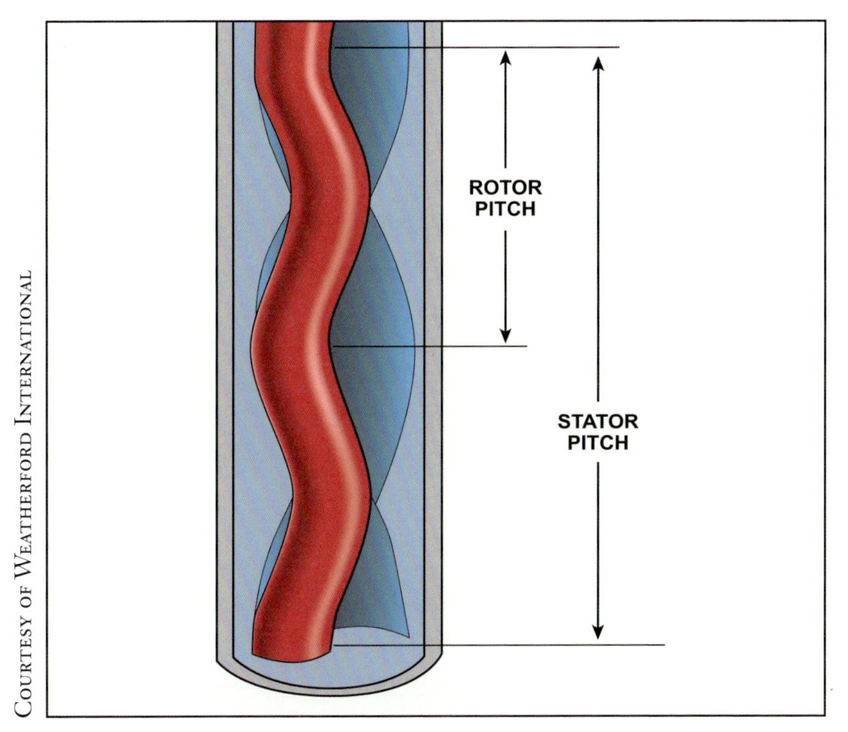

Figure 43. PCP rotor and stator pitches

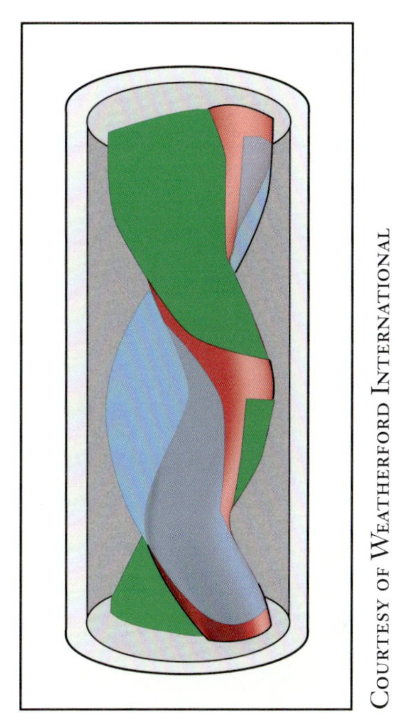

Figure 44. The rotor in a PC pump creates overlapping cavities.

hydraulic motors, or it can be belt driven by gas engines. The surface drive system must have a brake capable of dynamically dissipating the energy in the rod string by back-spinning the rods in a controlled manner in order to maintain operational safety.

While rotors are designed to control fluid fallback, some slippage is desirable because it helps lubricate and cool the pump.

Figure 45. 1:2 and 2:3 stator configurations

The high efficiency of the PC pump system is the result of positive-displacement rotary motion with no dead cycle time. The rod drive system is likewise efficient since it consists of direct-coupled mechanical components. The amount of slippage or fluid fallback, and therefore volumetric efficiency, can be controlled by rotor-to-stator fit. However, some slippage is necessary for lubrication and cooling to assure acceptable pump run life.

System Components

Tubing conveyed progressing cavity pumps have the stator threaded directly onto the bottom of the production tubing. The rotor is run into the well attached to the bottom of the rod string. It is positively positioned inside the stator by passing through the stator to a tag bar in the pump subassembly below. Then the rod string is retracted a calculated amount to account for anticipated rod stretch and thermal expansion so that the rotor is correctly positioned within the stator during pumping.

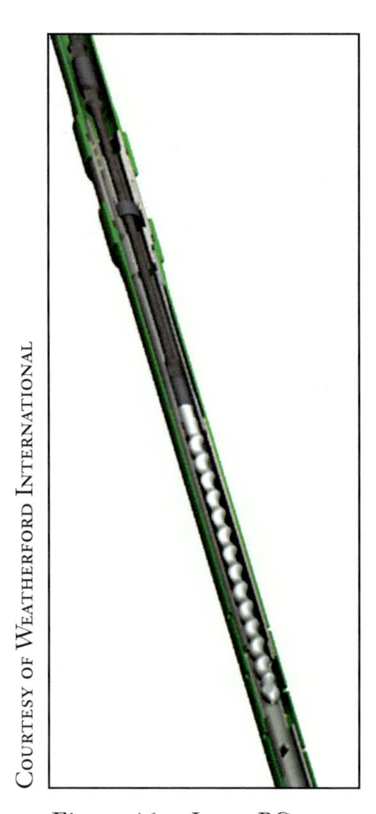

COURTESY OF WEATHERFORD INTERNATIONAL

Figure 46. Insert PC pump

Insertable progressing cavity pumps (or, insert pumps) are run into the well as complete units inside the production tubing on the end of the rod string, similar to the way insert rod pumps are run (fig. 46). A hold-down anchor assembly is positioned in the subsurface pump seating nipple. The entire insert pump can be removed by pulling the rods, making it easy to service. Because the entire insert pump must fit inside the production tubing, insert PC pumps typically have lower volumetric capacity due to their smaller diameter.

Rod strings for rotating applications can either be jointed sucker rod, continuous rod, or small diameter tubing (called, hollow rod). High-strength and large-diameter jointed sucker rods and hollow rods are used for higher-volume and deeper applications. Continuous rods are preferred for many heavy oil applications in order to eliminate flow restrictions associated with couplings. Continuous rods are also preferred for deviated well geometries to reduce failures due to rod/tubing contact wear from couplings. While most PC pump applications use conventional sucker rods or continuous rods, special sucker rod connections have been developed to improve rod performance for rotating rod applications. It is important to note that the rod makeup procedures for PC pump applications are different than for reciprocating rod applications.

Driveheads can be powered by electricity (fig. 47). Alternatively, they can be powered by gas engines where electricity is not available (fig. 48). Oilfield electric motors turn at 1,200 to 1,800 revolutions per minute, and gas engines run at similar speeds or higher.

Rod strings that are used for rotating applications can be:

- Jointed sucker rod
- Continuous rod
- Hollow rod

Courtesy of Weatherford International

Figure 47. Electric drivehead for a PC pump

Courtesy of Weatherford International

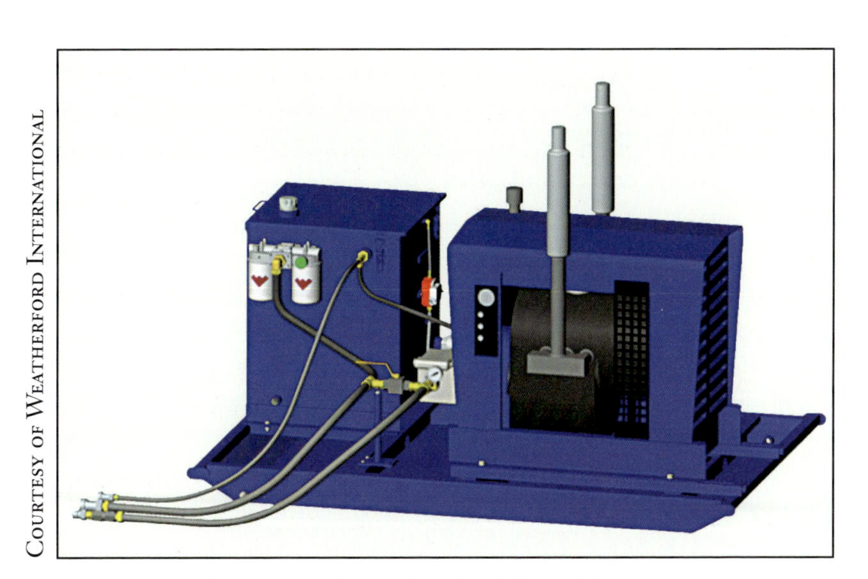

Figure 48. Gas engine-driven drivehead for a PC pump

Courtesy of Advantage Products Inc.

Figure 49. Torque anchor

Therefore, some type of mechanical reduction is required within the drivehead to rotate the rod strings at the most common PC pump speeds of 100 to 400 revolutions per minute. The most common drivehead configurations utilize belts and sheaves since speed can be adjusted by changing the sheave ratios. Gear reducers are also used when speed flexibility is less critical or where speeds can be adjusted by means of electric or hydraulic transmission control. Driveheads using integral low-speed, high-torque electric motors have also been developed.

Torque anchors, sometimes called no-turn tools, are used in high-torque applications to assure that the tubing does not become unscrewed due to the torque generated by the pump. The torque anchors have slips that prevent rotation of the tubing relative to the casing (fig. 49).

System Design

The geometry of a pump is determined by its pitch, eccentricity, and displacement.

PC pump systems are sensitive to fluid properties and the application environment because of the elastomeric stator. Therefore, the design process must begin with a thorough understanding of the application and all the fluids to which the pump will be exposed, including chemicals for corrosion and diluents. This usually requires testing fluid samples in order to properly select elastomers and determine how they will behave when exposed to those fluids. This step is critical to properly determine the amount of rotor-to-stator interference fit required to meet target production rates and expected run life. With the fluid properties and application conditions known, the design process can be organized into the following steps:

1. Pump geometry and elastomer selection
2. Production tubing sizing
3. Rod string selection
4. Drivehead and prime mover selection

The pitch, eccentricity, and displacement define the basic pump geometry. Pump displacement is chosen based on target production volumes and assumed volumetric efficiencies. Production volume capacity is expressed in terms of a fluid rate at a given pump speed[4], such as 32 M^3/100 rpm (revolutions per minute). Planned actual volumes are then based on the planned operating speed and the design volumetric efficiency. Operating speed is typically dependent upon viscosity whereby

[4] ISO 15136-1

pumps for water and low-viscosity oils might turn at speeds of 400 rpm or higher, but pumps for more viscous oils would turn at less than half of those speeds. Expected pump reliability is also a factor in determining operating speed, as stator elastomer fatigue is dependent on the number of revolutions or cycles imposed. A PC pump that operates at 200 rpm might last twice as long as one operated at 400 rpm.

Pump components are typically sized to achieve volumetric efficiencies of 75% to 90%. Pumps can be designed to higher volumetric efficiencies, but pump life might be compromised due to inadequate cooling and high rotor-to-stator contact pressure. Pump efficiencies are specified relative to pump performance at standardized test conditions. Low test volumetric efficiencies, including zero efficiency (100% slippage) and even negative efficiency[5], can be specified to accommodate anticipated elastomer swelling under subsurface operating conditions. For example, in figure 50, in order to achieve 70% volumetric efficiency under normal operating conditions, the pump was sized to have zero efficiency under standard test conditions for water.

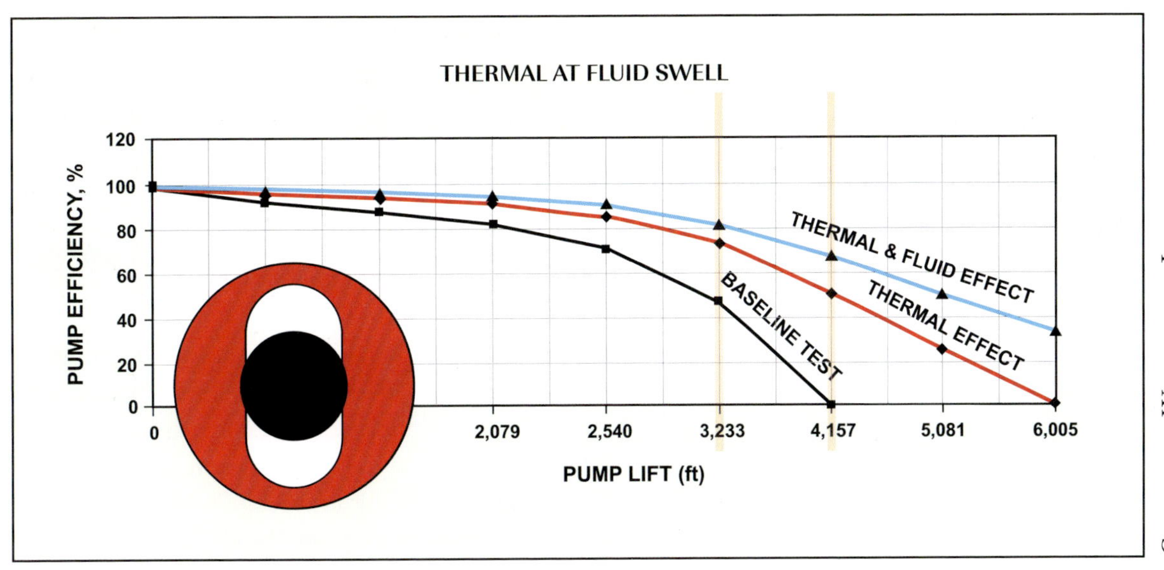

Courtesy of Weatherford International

Figure 50. Pump efficiency versus pump lift

[5] Volumetric efficiency ratings for PC pumps are based on standardized test conditions. Efficiency is varied by varying the relative interference between the pump rotor and stator. Negative efficiency is a term for specifying specific rotor-to-stator clearances rather than interference under test conditions. Low, zero, and negative efficiencies are specified to generate looser fitting pumps at standardized conditions in anticipation of elastomer swell in actual well conditions.

To determine the lift rating of a pump, multiply the incremental pressure capacity of individual cavities by the number of pitches.

The anticipated subsurface temperatures and well fluid chemistries will cause the stator to swell under normal operating conditions to achieve the desired 70% efficiency. For this reason, the production from newly installed PC pumps should gradually increase to target levels over a period of days or weeks as the elastomers swell to achieve the desired rotor-to-stator fit.

Pump lift is the pump pressure rating based on the pressure boost between the pump intake and the pump discharge, measured in depth (in feet or metres). The lift rating of a pump is determined by the incremental pressure capacity of the individual pump cavities multiplied by the number of pitches in the pump. The capacity for a pump to lift fluid must be adjusted for specific applications by considering flow losses, inefficiencies due to gas entrainment, flow-line back-pressure, boost from charge pumps, and similar factors. For this reason, a pump rated at 9,000 feet (2,743 metres) of lift might be required to lift fluids from 6,000 feet (1,829 metres) of true vertical depth.

The elastomers in PC pumps can be susceptible to explosive decompression in the presence of certain gases, such as CO_2. They are also susceptible to attack from aromatic gases such as benzene, xylene, and toluene, which limits their use in higher API oils. Aromatic gases and higher temperatures can swell elastomers, resulting in distorted stator geometry and improper rotor-to-stator fit.

Because PC pumps are sensitive to temperature, CO_2, aromatic gases, and the rotor-to-stator fit; the elastomer materials and pump dimensions must be designed for each specific application. Because the pump motion consists of a hard metallic surface sliding against an elastomer, the pump design must allow some fluid slippage for lubrication and heat dissipation. In addition, a pump cannot be operated in a pumped-off condition and allowed to run without fluid production; if this occurs, the elastomer will quickly overheat and fail. Therefore, although PC pumps often represent the most efficient lift technology, they also represent the least forgiving lift technology if well conditions and pump design are not well understood and matched properly.

For tubing conveyed PC pump installations, the production tubing should be sufficiently large to minimize flow losses and to accommodate the pump rotor geometry. The inside diameter of the tubing must be sufficiently large for the rotor's outside diameter to pass inside when the rotor is run into the well using the sucker rod.

The tubing must also accommodate the eccentric orbit of the rotor head during pumping operations. In applications with sandy fluids, the tubing size should be minimized in order to keep flow velocities sufficiently high to keep the sand in solution. This will prevent sand fallback and bridging.

Tubing size should be minimized in sandy fluids to maintain high flow velocities and prevent fallback and bridging.

For insert PC pump applications, the diameter inside the tubing must also accommodate the subsurface pump assembly. Most insert installations rely on a landing nipple to locate the pump, but configurations without nipples are also possible.

The rod string is sized based on combined stresses from torque and axial loads. Hydraulic torque is the result of the pump displacement and lift. The frictional torque of the pump is caused by:

- Sliding/rolling interaction between the rotor and stator
- Hysteretic losses associated with the deformation of the elastomer
- Fluid losses through the pump

The rod string will experience a fluid drag resistance torque as it turns in viscous fluids. The rod string will also have an axial load from the weight of the rod string, the pump axial thrust resulting from the discharge pressure acting on the pump-load bearing area, and any surface back-pressure.

As with reciprocating rod systems, jointed rod strings in PC pump systems sometimes cause tubing wear due to contact between the rod couplings and the tubing. The severity of the tubing wear is dependent upon wellbore curvature and deviation, sand, pump speed, fluid lubricity, and similar factors. Tubing rotators can be used to slowly rotate the tubing, thus spreading out the wear and delaying tubing failure. Also, the rods can be periodically raised or lowered to spread out the contact. Rotary sucker rod couplings and rod centralizers can reduce tubing wear. Continuous sucker rod or hollow rod with flush connections is ultimately the most effective solution, as the rod side loads are spread out over the length of the rod rather than being concentrated at the couplings.

Driveheads are chosen based on the type of energy available in the field. Electric motors are used if electricity is available and reliable; otherwise, gas engines are used. Sizing is based on the required bearing to support the axial load of the fluid column, plus the torque and speed required for the application.

Installation and Operating Considerations

The system installation should include a flow tee, a *blowout preventer* (*BOP*) that closes on the polished rod, and a means (other than the drivehead) of holding both the torque and the axial weight of the sucker rod string, so that the drivehead and stuffing box can be serviced without a workover unit. Composite BOPs are available that act as flow tees and which can hold polished rod weight and pump torque (fig. 51). These composite BOPs are recommended for all PC pump installations.

Operating speeds should be within the range recommended by the pump supplier. Excessive speed can cause rod whirl, which can result in damage to the rod system and tubing. At very low speeds, the system can experience stick-slip behavior. During stick-slip, the pump alternately stops rotating and then rotates rapidly as the surface drivehead continues to rotate at constant speed. Stick-slip can damage the pump, rods, and drivehead.

COURTESY OF WEATHERFORD INTERNATIONAL

Figure 51. Composite flow tee with rod-gripping feature

Summary

Progressing cavity pumps (PC pumps) consist of a rotor that is placed inside a helical cavity stator. This assembly creates a series of sealed cavities that move upward as the rotor turns. Conventional rod-driven PC pumps are rotated by rod strings from a surface drivehead. They are used primarily for land applications. PC pumps driven by ESP-style electric motors (ESPCPs) can be used offshore and in deviated wells. PC pumps are used where operating efficiency and low initial capital costs are critical. They are preferred for producing viscous fluids, such as cold, heavy oils, as well as abrasive fluids containing particulate matter. They have also proven effective for pumping water and medium to light oils. PC pump systems are sensitive to temperature, CO_2, aromatic gases, and the rotor-to-stator fit. Because certain chemicals can cause corrosion and elastomer swelling and shrinkage, the design of a PC pump system must begin with a thorough understanding of the fluids to which the pump will be exposed.

Conventional Gas Lift

In this chapter:

- Typical applications of conventional gas-lift systems
- Distinctions between continuous and intermittent gas lift
- Key factors to consider when using gas lift
- Achieving maximum efficiency with a gas-lift system

Air lifting of water with a small amount of oil was first known to be used in the United States as early as 1846, but compressed air was reportedly used to lift water from wells in Germany as early as the eighteenth century. These systems operated initially in a very simple manner by induction of air to the bottom of the tubing and out into the casing. *Aeration* of the fluid in the casing-tubing annulus decreased the weight of the mixture to the extent that fluid would rise to the surface and flow out of the well. The process was sometimes reversed by injecting down the casing and producing through the tubing.

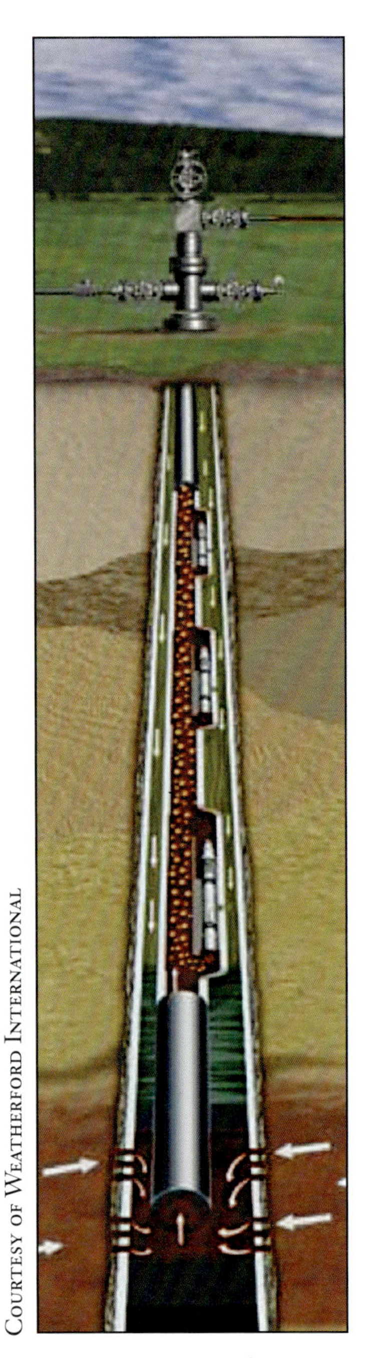

Courtesy of Weatherford International

Figure 52. Gas lift

Operators continued to use air lift as a means for lifting oil during the early part of the twentieth century. By the 1920s, natural gas became widely available as an option for lifting fluid. Since gas is lighter than air, it provided operators with a high-performing lifting alternative. Also, gas lift reduced the rate of damage caused by *oxidation*, which frequently deteriorated well equipment. During the 1930s, several types of gas-lift valves were developed that offered better gas injection and replaceable wear components. Gas-lift valves using pressure operation principles date back to the King valve, patented in 1944, and numerous bellows-operated valves have been developed since then. In 1954, Harold McGowen and Howard Moore patented the wireline retrievable valve which simplified gas-lift valve replacement and made gas lift a competitive method of production, especially where gas at adequate pressure was available for lift purposes.

Today, gas lift covers a variety of practices by which gas is injected into the tubing to increase the production of a well or to restore production where the well is dead (fig. 52). Injection can be accomplished by a simple perforation in the tubing, a jet collar in the tubing, or gas-lift valves designed and preset to open or close under specific well conditions. The injection point can be at any depth from a few hundred feet to twelve thousand feet or more depending on the application requirements.

Continuous Gas Lift

Continuous gas lift artificially increases a well's gas-liquid ratio to reduce the density of the fluid column in a flowing well so that less formation pressure is required to lift the fluids to the surface. Unlike many other artificial-lift technologies, formation gas in the fluid stream improves gas-lift pumping efficiency rather than compromising efficiency. Conventional gas lift is not negatively impacted by sand and particulate matter. It can lift effectively in both deviated and vertical applications.

Typical Applications

Continuous gas lift is most applicable to wells in reservoirs with substantial amounts of remaining energy and high productivity indexes. This lift method works best as an assist to a reservoir capable of sustaining high fluid levels resulting from sustainable moderate formation pressures at target production levels. For this reason, continuous gas lift works well in water drive reservoirs where high reservoir pressures are maintained throughout the life of the well. Continuous gas lift becomes less attractive as the volumes lifted decrease and as formation pressures decline.

Continuous gas lift is equally effective in land and offshore applications but is particularly ideal for offshore applications as there are no rods, cables, or moving parts to interfere with subsurface safety valves. Similarly, gas lift is equally effective in vertical and deviated well geometries, although the lifting primarily occurs in the more vertical section of the well. Gas-lift valves can be run and retrieved in well sections with up to 60° deviation using conventional *wireline* tools and in sections with over 70° deviation using special wireline tools; the latter configuration can include special rollers to reduce sliding friction.

Gas-lift systems are effective for a number of applications, including:

- Land
- Offshore
- Vertical wells
- Deviated wells

Operating Principles

Continuous gas lift is a form of *natural flow*. With natural flow, the energy of compressed gas in the reservoir is the principal force that raises the well fluids to the surface by pressure and aeration:

- Pressure of the gas exerted against the fluid at the bottom of the tubing might be sufficient to lift the entire column of fluid to the surface.
- Aeration of the column of fluid by gas bubbles entering it at the bottom of the tubing reduces the density of the column of fluid. As the gas moves up the tubing, each bubble of gas expands because of the reduction of pressure, further reducing the density of the fluid column.

Natural flow in the well continues as long as the formation pressure can lift the well fluids. As pressure declines over time, the reservoir pressure alone will no longer be sufficient to lift fluids to the surface. An increase in the percentage of water can also cause an oilwell to stop flowing. Because water is denser than oil, it increases the weight of the fluid column. Furthermore, because water does not contain gas in solution, there is less naturally occurring gas in the tubing to reduce the density of the fluid column. When natural flow ceases, the well fluids rise to an equilibrium static level in the well and must be lifted to the surface by artificial means.

Continuous gas lift provides additional gas to increase the aeration of the well fluids in the production tubing (fig. 53). Pressurized gas is pumped into the annulus between the casing and the tubing above a packer. The gas displaces any liquid in the annulus through the gas-lift valve(s) into the production tubing string. The gas is then injected through one or more gas-lift valves into the production tubing where it expands to aerate the fluid inside. The resulting decrease in fluid density and decrease in fluid column weight cause the formation pressure to lift the fluids toward the surface. As the gas moves up the tubing and the hydrostatic pressure decreases, the gas expands to further decrease the weight of the fluid column.

Continuous gas lift does not provide additional lift pressure beyond what is available in the formation to lift the fluids. Excessive gas injection pressure can increase the mass flow rate in the production tubing, increasing friction and causing an increase in the flowing pressure at the bottom of the well. Therefore, injection pressure must

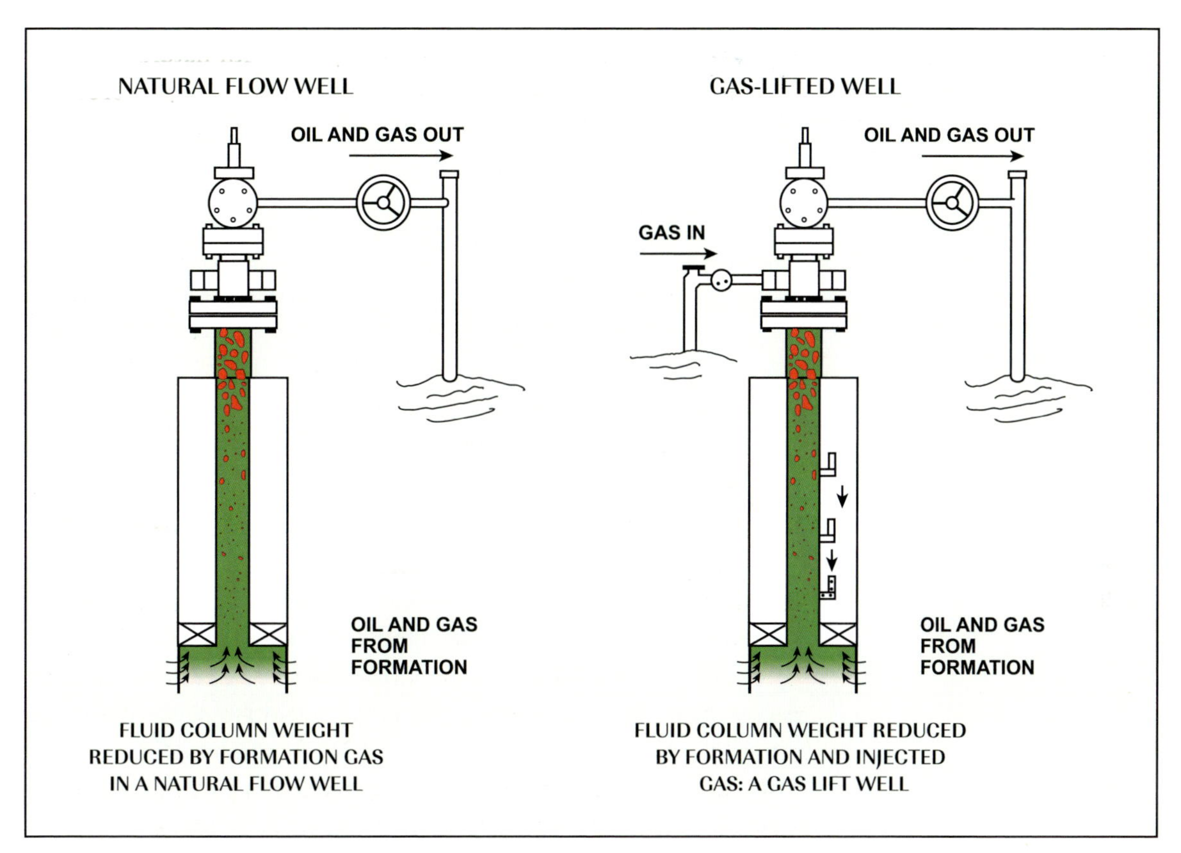

Figure 53. Natural flow and gas lift

be sufficient to overcome the static fluid pressure in the production tubing at the point of injection without exceeding the formation pressure that is driving the fluids through the perforations into the well. Accordingly, the amount of injected gas must be continuously metered to provide sufficient aeration for the formation pressure to lift the fluids to the surface without overlifting the well.

Maximum lift efficiency is achieved when the gas is injected as low in the tubing as possible so that the maximum amount of well fluid becomes aerated. Therefore, the bottom gas-lift valve should be set as low as possible in the tubing string, and during steady state operations, all of the lift gas should be injected through this valve alone.

Gas-lift systems usually have multiple injection points in order to reduce the pressure required to initially unload the well of liquids.

To initiate gas lift, the well must first be unloaded by removing any liquid in the annulus above the lowest gas-lift valve so that gas can be injected through that valve. Unloading is achieved by pressurizing the annulus with injection gas to force the annulus fluid through the open gas-lift valve(s) and into the production tubing.

Although gas injection through one valve is optimum, unloading through one deep gas-lift valve can create challenges. Substantial injection pressure would be required to displace all of the annulus fluid and gas through a single deep gas-lift valve when the production tubing is full of liquid. The costs of high-pressure compression and associated high-pressure flow lines and wellheads are often prohibitive. Therefore, gas-lift installations often consist of multiple injection points to sequentially lighten sections of the fluid column in the production tubing, thereby reducing the required unloading pressure.

Unloading the well using multiple gas-lift valves occurs in stages. In the most common gas-lift configuration, initially all of the gas-lift valves that are not covered by liquid are closed, and those covered by liquid are open due to hydrostatic pressure. When injection gas enters the annulus, it displaces annulus fluids through the open gas-lift valves and into the production tubing. As the liquid level in the annulus drops below the upper gas-lift valve, injection gas begins to enter the tubing through the upper gas-lift valve to aerate the fluid in the tubing above the point of injection. The fluid level in the annulus continues to drop as annulus fluid flows through the open gas-lift valves into the production tubing due to continued gas injected into the annulus or reduced fluid pressure in the tubing resulting from fluid aeration. When the liquid level in the annulus reaches the second open gas-lift valve, gas begins to be injected from that valve to aerate the fluid in the tubing from that point to the surface. As the annulus fluid level drops, each valve in turn is uncovered and begins injecting gas into the tubing. At the same time, the upper valves begin to close in sequence in response to the changes in pressure. The unloading sequence continues until the bottom valve begins to inject gas and all upper valves close. In this way, the final pressure required to incrementally aerate the production fluids, unload the well, and inject through the deepest gas-lift valve is much less than the pressure that would have been required for unloading a column of 100% liquid through a single deep valve. The quantity, sizing, and spacing of gas-lift valves must be precisely determined based upon the available gas injection pressure, well depth, and other factors in order for gas lift to be effective (fig. 54).

The quantity, size, and spacing of gas lift-valves are determined by a number of factors, including the available gas injection pressure and well depth.

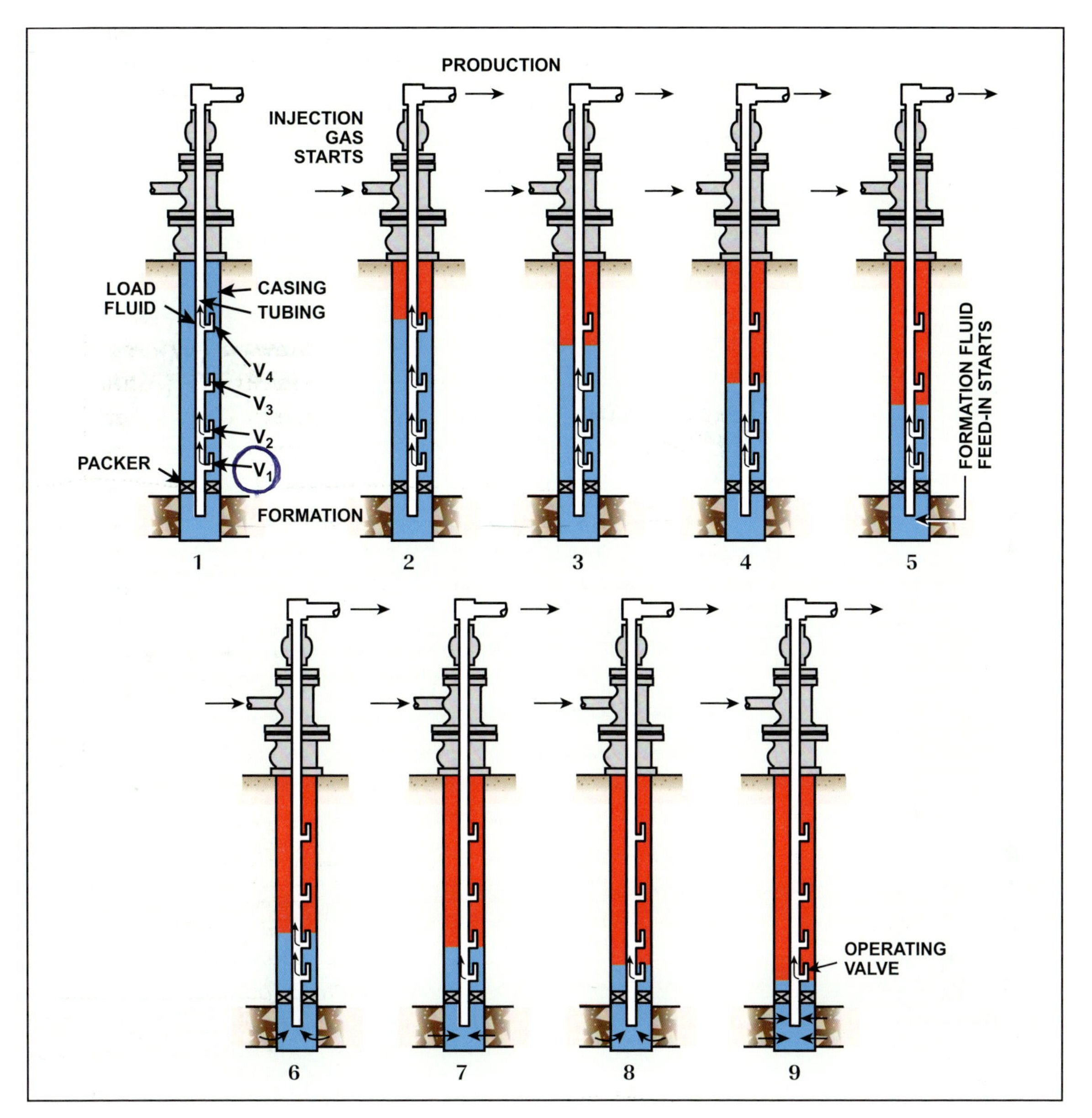

Figure 54. Unloading sequence

During the unloading process, the open valves share the flow of liquid from the annulus into the production tubing, reducing the volume of annulus fluid to be unloaded through the bottom gas-lift valve. This flow sharing reduces the volume and velocity of liquids through each valve, thus reducing the potential for erosion of the flow ports in the gas-lift valves.

The use of multiple gas lift valves for unloading helps prevent erosion of the main producing valve.

> Use of one deep gas lift valve eliminates potential failure points, but initial unloading flow rates must be limited to avoid eroding the valve.

Single gas-lift valve systems can be used where high-pressure injection gas is available. In many subsea applications, single-point gas lift can be used because high-pressure gas compression is already required to overcome additional back-pressure due to flow resistance through long subsea transmission lines. However, this flow-induced, back-pressure is not significant during the lower flow rates of the unloading process. As a result, all of the high-pressure gas is available for the unloading process, so unloading through a single deep gas-lift valve is possible. The use of a single gas-lift valve reduces the number of potential failure points in the system, but care must be taken not to erode the valve by unloading liquids at excessively high rates.

System Components

Gas-lift valves are most often designed to be normally closed until certain conditions of pressure in the annulus and the tubing cause the valves to open. When a gas-lift valve opens, it permits gas or fluid to pass from the casing annulus into the tubing. Alternatively, gas-lift valves can be arranged to permit flow from the tubing to the annulus, although this is not common.

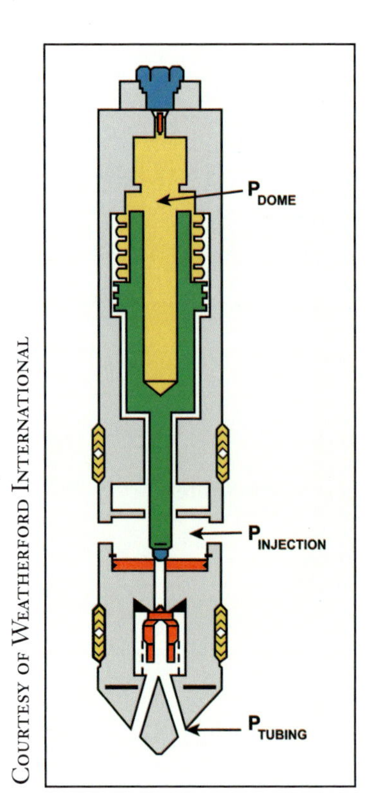

Figure 55. Bellows gas-lift valve

Figure 55 illustrates the basic operating principles of a bellows-type gas-lift valve. Mechanisms used to apply force to keep the valve closed include a dome or chamber charged with a gas, usually nitrogen, under pressure; or a spring in compression. Each of these is adjustable at the surface before the valve is run into the well so that the valve closes when the target pressure condition exists. The dome or chamber pressure setting is adjusted for the expected well conditions. Various combinations of these two mechanisms will be found in most gas-lift valves.

All gas-lift valves are constructed for one-way flow—that is, a check valve function is included in the assembly. Ordinarily, the check valve is arranged to allow flow from the annulus to the tubing. When a packer is used below the gas-lift valve to seal the annulus, after initially unloading the annulus the check valves keep the annulus unloaded above the packer. This makes it possible to shut down and restart the well without having to repeat the unloading process. The check valve also protects the casing by keeping well fluids off the casing above the packer.

Today, gas-lift valves are available in several basic configurations and are selected depending upon the requirements of each application:

- IPO: Normally closed, opens when injection pressure reaches a preset pressure.

- PPO: Normally closed, opens when the production pressure reaches a preset pressure.
- Orifice: No closing mechanism, always open. Commonly used as the lowest injection point.
- Pilot Operated: Normally closed, rapid opening of large orifice by an internal pilot section. Used for intermittent lift.
- Balanced: Controlled by both injection pressure and production pressure; applicable in special applications such as dual gas-lift installations; rarely used.
- Dummy: No opening, always closed.

When paired with an integrated field management system, a rate controller can be used remotely to adjust injection rates.

Gas-lift valves are held in place by gas-lift mandrels in the production tubing string. These mandrels can be either conventional or side-pocket types. A conventional gas-lift mandrel has the valve mounted on the outside of the tubing. To replace the gas-lift valves, the tubing must be pulled from the well. A side pocket mandrel has an internal pocket that holds the gas-lift valve and a communication opening to allow fluids to pass from the annulus through the gas-lift valve and into the production tubing (fig. 56). The valves are run into the well and landed in the mandrel pockets by means of special running tools, typically suspended on wireline. Likewise, running tools

Courtesy of Weatherford International

Figure 56. Gas-lift mandrel

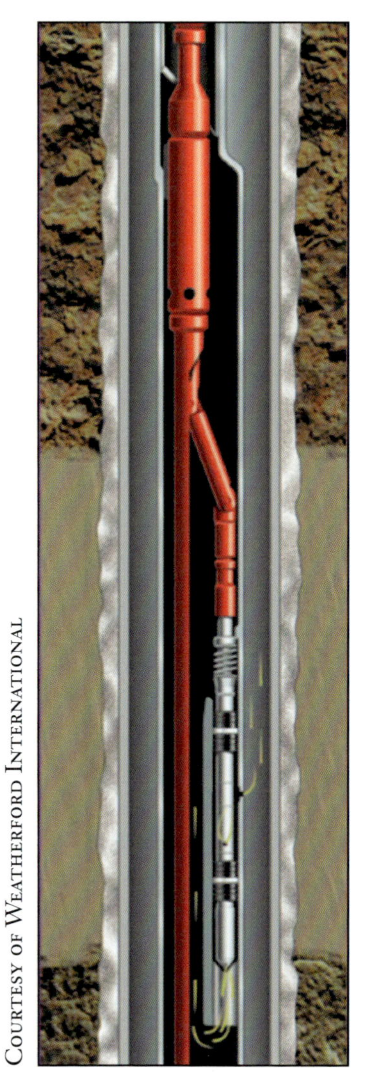
Courtesy of Weatherford International

Figure 57. Gas-lift running tool

can be lowered to latch and retrieve the gas-lift valves. Therefore, gas-lift valves can be easily replaced or changed as needed (fig. 57).

One or more devices are installed on the surface to control the timing and amount of gas injected. A fixed choke is used where adjustment of injection pressure is to be minimized or when it is used in series with other control devices. Adjustable choke valves, typically needle valves, provide a simple manual means to adjust injection rates. A rate controller with a surface motor valve can be used for remotely adjusting injection rates and for operation in conjunction with an integrated field management system.

The produced fluids are discharged into a conventional fluid treatment system, which usually includes an oil and gas separator. Flow restrictions in the surface connections should be minimized, and the back-pressure on the separator should be kept as low as possible. Tubing and casing pressures must be monitored, and the amount of injection gas metered accordingly to optimize the lift performance and gas usage.

System Design

Additional gas can be injected through perforations in the tubing, helping to lift fluids when reservoir pressure has declined. However, excessive gas will eventually limit efficiency.

The best candidates for continuous-flow gas-lift applications are wells completed in reservoirs with substantial amounts of remaining energy and having high productivity indexes. Wells of this type are usually located in reservoirs with good or moderate natural water drives where reasonably high production fluid levels in the wells can be expected during much of the producing life of the reservoir.

When water production increases or reservoir pressure declines, a well might stop flowing. Instead of terminating the well to install gas-lift valves at this stage of depletion, an operator may elect to *perforate* a small hole in the tubing, thus permitting injection of high-pressure gas to lift the well fluid (fig. 58). The depth to which this perforation is made in the tubing must be based on the flowing pressure gradient of the well, static fluid level, available gas pressure, and desired rate of production. Additional perforations at selected lower depths might be helpful later as the producing rate decreases, but eventually the added perforations will make the gas lift very inefficient due to the excessive gas injected through the upper tubing perforations.

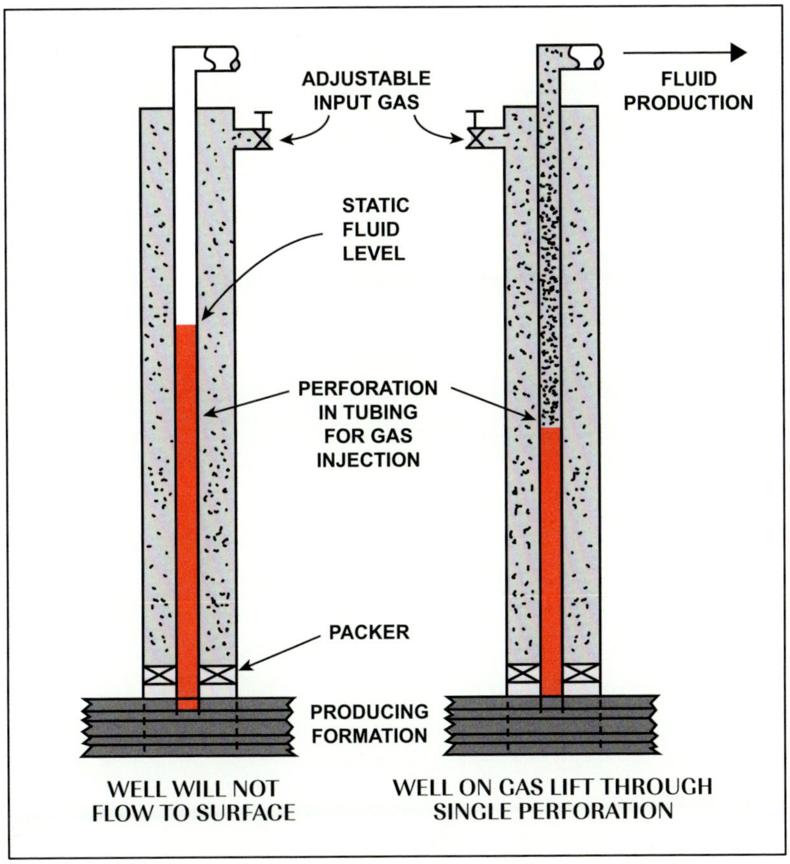

Figure 58. Single perforation for gas lift

A gas-lift configuration can be:

- Open installation
- Semi-closed installation
- Closed installation

The well's producing rate might actually decline upon addition of perforated holes in the tubing because of excessive back-pressure caused by the induction of too much gas into the tubing. At this point, installation of gas-lift valves will improve production and improve the efficiency of gas usage. Here again the productivity index, bottomhole pressure, available gas pressure, and desired flow rate must be known so that the appropriate gas-lift valves can be selected and installed at proper depths in the well.

The simplest gas-lift configuration is an open installation, which does not include a packer or standing valve. Casing pressure will act against the formation. Fast unloading is possible since the fluid is transferred into the bottom of the tubing. Each time the well is shut in, the formation fluid will build up in the annulus, requiring the well to be unloaded when put back into production.

A semiclosed installation includes a packer but no standing valve. The packer seals off the annulus from the production fluids and prevents injection gas from blowing around the bottom of the tubing. The packer and gas-lift valves will prevent the unloaded annulus from subsequently filling with well fluids. Once the annulus is unloaded, the working level of the fluid in the annulus remains stable, insuring faster restarting of the production after temporary shut-in. The initial unloading of the annulus is slower because all the fluid must be transferred through the gas-lift valves in the tubing string.

A closed installation utilizes a packer and a standing valve. The packer seals off the annulus, and the standing valve permits flow in only one direction. Standing valves are used for continuous gas lift in low-production wells to help retain produced fluids from falling back as pressures fluctuate. Intermittent gas lift should be considered as an alternative to closed installation continuous gas lift if the field gas supply will tolerate the intermittent gas demand cycle.

Selection of the type of equipment for continuous flow should be made by taking into consideration the well and reservoir characteristics and the existing equipment in the well, as follows:

- Depth of well to producing formation
- Size of tubing in the well
- Size of production casing
- Maximum gas pressure available for kickoff
- Volume of gas available
- Gas pressure available for continuous flow after kickoff
- Daily volume of fluid to be produced and percentage of water
- Static well pressure gradient
- Flowing well pressure gradient and gas-fluid ratio
- Back-pressure expected from surface equipment

Intermittent Gas Lift

Intermittent gas lift is similar to continuous gas lift in that it uses gas to lift fluids. It shares much of the same equipment as continuous gas lift, and it typically involves releasing pressurized gas from the annulus into the production tubing. However, intermittent gas lift is designed to lift slugs of liquid by sudden release of high velocity bursts of pressurized gas rather than by continuous aeration of the production fluid.

Typical Applications

Intermittent gas lift—

- Most often applies to oilwells with low productivity
- Is also used for dewatering gas wells where injection gas is available

Intermittent gas lift is most often applied to oilwells with low-productivity indexes such that accumulation of fluids in the bottom of the well will take place over a fairly long period of time. It can also be used for dewatering gas wells where injection gas is readily available. Intermittent lift does not use the reservoir gas energy, so it is less efficient than continuous gas lift. Intermittent gas lift creates an intermittent gas demand on the injection gas supply system, and the well fluids and gas arrive in slugs that require high-capacity separation equipment. In fields with numerous intermittent gas-lift wells, the spikes in gas demand and produced fluid volumes tend to smooth out due to aggregation of supply gas and fluid processing for the field. Intermittent gas injection is undesirable for unconsolidated formations that produce sand because the sand is allowed to settle in the tubing between pump cycles.

Operating Principles

Formation energy raises the fluid to the static level in the well. After a set time interval, high-pressure gas is released into the casing. At a preset pressure, an injection valve snaps open and releases gas into the production tubing at a high rate. The fluid slug above the injection point is lifted by the gas at a rapid rate to minimize fluid fallback. As the slug is propelled toward the surface, the injected gas expands to further accelerate the liquid slug. The gas is then shut off, and fluid is allowed to build up until the next intermittent cycle. A time cycle controller and choke or motor valve control the frequency and duration of the gas injection.

As with continuous gas lift, multiple gas-lift valves can be installed above the operating valve to initially unload the fluid from the annulus. The packer and gas-lift valves will prevent the unloaded annulus from subsequently filling with well fluids. With the liquid removed from the annulus at the upper valves, the upper valves will remain closed during gas injection cycles. With each injection cycle, all gas is injected through the lowest gas-lift valve.

System Components

Much of the same equipment used in continuous gas lift is used in intermittent gas lift, including mandrels and surface supply gas and well fluid processing systems. Special pilot-operated, gas-lift valves are used to release sudden slugs of gas into the production tubing. An internal pilot section in the valve fully opens it when the pressure setting of the valve is reached. This allows the annulus to accumulate a substantial amount of compressed gas energy before sudden release of the gas into the production tubing. Controllers are used to set the injection cycle frequency and duration period.

System Design

The goal of intermittent gas lift is to maximize the amount of liquid produced while minimizing the amount of injection gas used. The design of the intermittent gas-lift system will be dependent upon the productivity index (PI) and permeability of the well, the anticipated produced fluid rate, the well construction, and the surface facilities, including the availability of pressurized gas.

The injection cycle frequency should maximize production while minimizing injection gas usage.

A semiclosed installation includes a packer but no standing valve as previously discussed under Continuous Gas Lift. The packer protects the casing annulus from well fluids and eliminates varying fluid levels in the casing, but the tubing pressure—including the injection pressure—is exerted against the formation. Thus, injection gas pressure can drive fluids back into the formation. This type of installation is suitable for high- to medium-static bottomhole pressures and a low-permeability formation with low-productivity index. IPO valves are generally used.

In closed installations, a packer seals off the annulus and a standing check valve isolates the injection pressure from the formation. Closed installations are necessary for high- to medium-productivity index wells to prevent injection gas pressure from pushing fluid back into the formation. Figure 59 illustrates a closed-type installation; note the standing valve in the tubing. The same well information required to design a continuous-flow installation is required to design an intermittent gas-lift installation. The uppermost valve is located at a depth below the normal static fluid level, which will permit kicking off the well with the pressure available from the gas-lift gas supply. The deeper valves are spaced according to the flowing pressure gradient in the tubing below the depth of the top valve. The number of unloading valves required will vary with the flowing pressure gradient and depth of the producing formation, assuming that the well will be intermitted from total depth. A time cycle inlet gas controller is installed at the well to control slugging of gas into the casing and the opening and closing of the operating valve (or intermitting valve) located in the tubing string just above the packer.

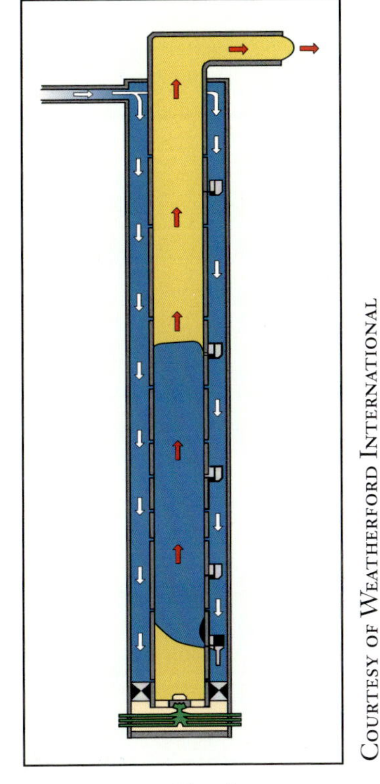

Courtesy of Weatherford International

Figure 59. Closed configuration, intermittent gas lift

It is important to tune the intermittent lift injection cycle frequency and injection period (injection duration) for each individual well. Cycle frequency and gas injection period should be set to optimize production while minimizing injection gas. Additional liquid can be produced by increasing the injection period, but once the liquids are lifted, continued gas injection is wasted. Also, the injection cycle frequency impacts liquid production. Cycles that are too long between injection periods leave the well loaded with liquid longer so that further inflow is deferred. Cycles that are too short result in a higher percentage of cycle time, as the system pressure bleeds down after the injection period.

Excessive gas injection can decrease liquid production due to increased flow friction.

This bleed-down time is dependent on how quickly the surface facilities can absorb the injection pressure pulse from the well.

In the example in table 3, the injection frequency and the injection duration were varied to find the optimum injection cycle and injection period. Notice that the well was capable of producing just over 28 barrels of liquid per day. This was accomplished with 34 seconds of injection time on a one-hour cycle. Extending the injection duration did not increase the amount of fluid produced, presumably because the well became effectively unloaded within 34 seconds. Any additional injection gas would be wasted. Reducing the injection duration below 34 seconds would not allow enough time to lift the entire fluid slug to the surface. Increasing the cycle rate from one hour to two hours reduced inflow of liquid and formation gas because the well was loaded for a longer period of time. Thus, less liquid and formation gas were produced. Reducing the cycle rate from one hour to 30 minutes decreased the amount of liquid lifted because much of the cycle time was consumed by the well bleeding down before additional inflow could occur.

Figure 60 shows the relationship between the gas injection rate and the produced liquid rate for the intermittent gas lift example in table 3. The diagram shows the amount of liquid that will be delivered relative to the amount of gas injected. Notice that there is a point at which additional injected gas no longer increases liquid production. Moreover, additional gas injection above this amount will begin to decrease the amount of liquid lifted.

Table 3
EXAMPLE OF TEST RESULTS WITH VARIOUS TIME CYCLES AND GAS INJECTION PERIODS

Injection Cycle	Injection Period	Total Fluid bbl per Day	Gas-Fluid Ratio per bbl
30 minutes	62 seconds	23.39	4,720 ft^3
30 minutes	46 seconds	20.31	4,280 ft^3
30 minutes	33 seconds	20.31	3,830 ft^3
30 minutes	21 seconds	16.23	1,928 ft^3
1 hour	64 seconds	28.41	2,042 ft^3
1 hour	44 seconds	28.53	1,533 ft^3
1 hour	34 seconds	28.44	1,110 ft^3
1 hour	26 seconds	12.18	2,035 ft^3
2 hours	60 seconds	20.31	1,400 ft^3
2 hours	46 seconds	20.40	1,082 ft^3
2 hours	37 seconds	16.23	1,084 ft^3
2 hours	31 seconds	16.23	938 ft^3

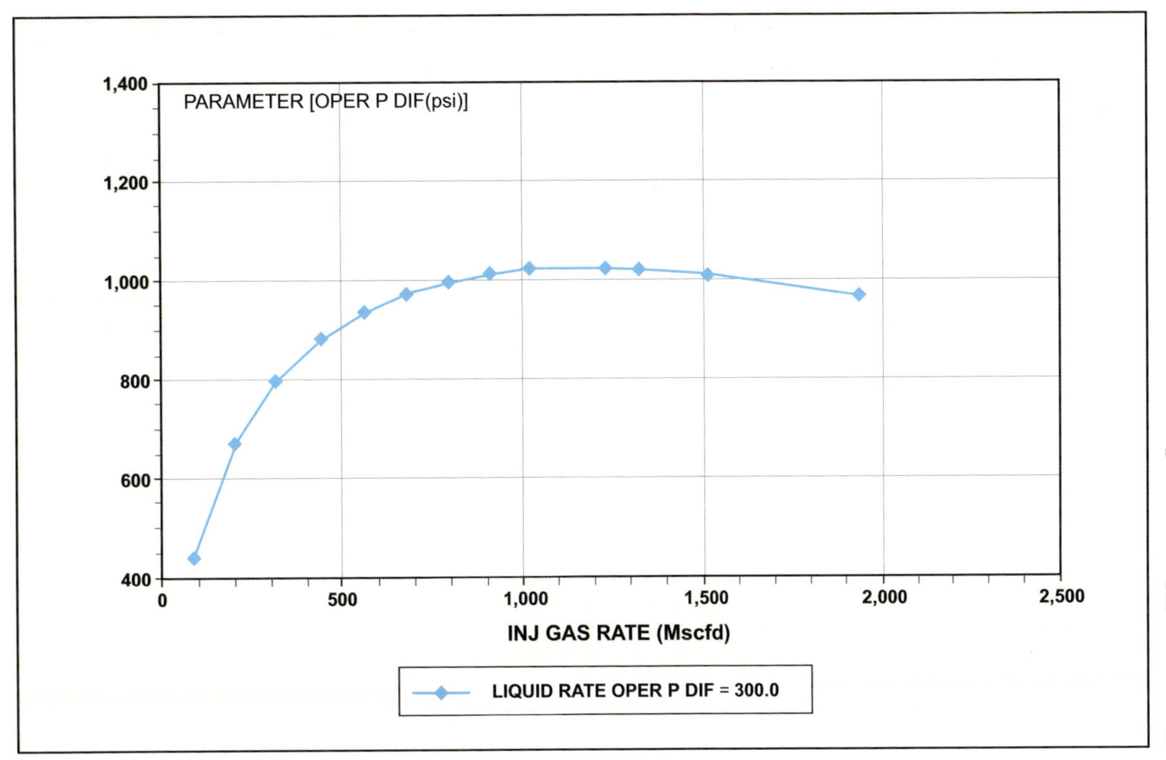

Figure 60. System deliverability curve

COURTESY OF WEATHERFORD INTERNATIONAL

Summary

In a gas-lift system, gas is injected into the tubing to reduce the density of the fluid column. This increases the flow of liquids to the surface. Gas lift is flexible. It is not negatively impacted by sand and particulate matter. It is effective in deviated and vertical wells. In contrast to many other artificial-lift technologies, the presence of formation gas in the fluid stream improves the efficiency of gas lift. Gas lift is equally effective in land and offshore applications. Maximum lift efficiency is achieved when gas is injected as low in the tubing as possible so that the maximum amount of well fluid becomes aerated. Therefore, the bottom gas-lift valve should be set as low as possible in the tubing string. Although gas injection through one valve is optimum, unloading through one deep gas-lift valve can create challenges. Substantial injection pressure would be required to displace all of the annulus fluid and gas through a single deep gas-lift valve when the production tubing is full of liquid. When multiple gas-lift valves are used to unload the well, the valves are opened in a sequence, starting with the uppermost valve, and then opening each lower valve in succession. One or more control devices are installed on the surface to control the timing and amount of gas injected. Gas-lift systems are configured as either open installation, semiclosed installation, or closed installation. The design depends on various factors of the well and reservoir. Intermittent gas lift uses gas released in bursts to lift the fluid column.

Plunger Lift

In this chapter:

- Typical applications of plunger lift
- Operating principles and functionality
- System components and their effectiveness
- Factors to consider when designing a plunger-lift system

Plunger lift is a method of lifting fluid by produced gas to drive a free-piston (plunger) from the lower end of the tubing string to the surface. This is done to remove accumulated fluid from the tubing string. Plunger lift is similar to intermittent gas lift in that it uses stored gas energy from the annulus or wellbore to periodically lift slugs of liquid, rather than lifting the entire column of fluid all at once.

The plunger lift overcomes two of the efficiency challenges of intermittent gas lift. First, the plunger acts as a mechanical interface seal between the slug of liquid that is lifted and the gas that moves the plunger and liquid. Thus, fluid fallback is greatly reduced, resulting in improved lifting efficiency. Second, the fluid is lifted using the energy of the formation rather than requiring pressurized injection gas energy from the surface. The result is the most cost efficient lift technology for low-volume applications (fig. 61).

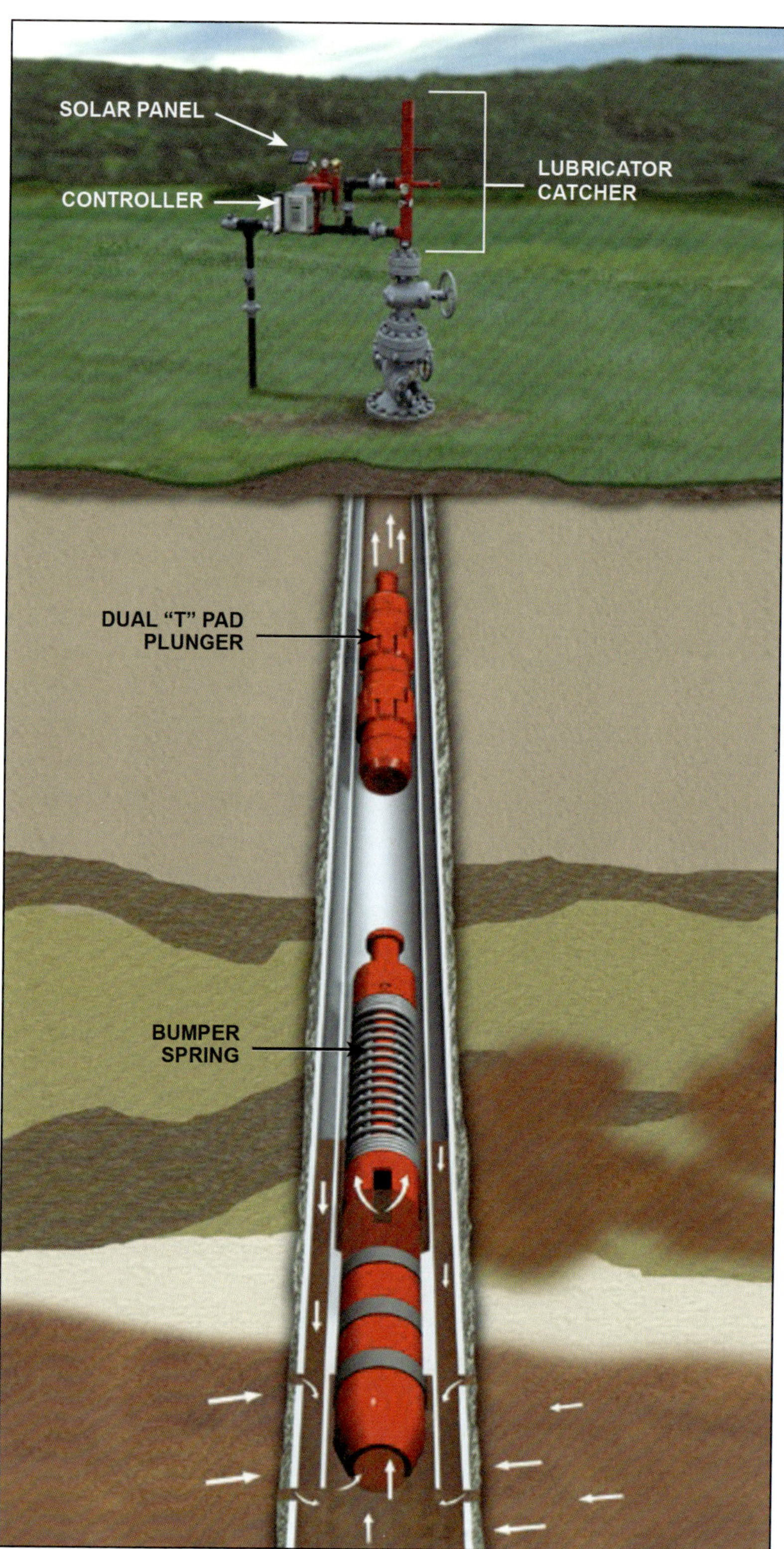

COURTESY OF WEATHERFORD INTERNATIONAL

Figure 61. Plunger lift

Typical Applications

Plunger lift is the lowest cost technology for removing fluids from gas wells that produce less than 200 barrels of fluid per day. It is most commonly applied to wells producing 1 to 50 barrels per day. Generally, plunger-lift systems can lift 100 barrels of liquid for each MMcf/d of produced gas. The installed equipment is relatively inexpensive, and there are no operating costs beyond minimal, periodic maintenance.

Plunger lift is also used to a lesser extent in oilwells, either as a stand-alone system in a gaseous oilwell or in conjunction with intermittent gas lift when formation gas is inadequate to lift the liquid slugs. When used with intermittent gas lift, the gas-lift-assisted plunger acts as a seal to prevent fluid fallback, thus improving efficiency. The injection gas also provides enough energy for the plunger to effectively remove scale and paraffin in the wellbore.

Operating Principles

Production is usually increased by maximizing the frequency of unloading cycles.

During normal production operations, the flow of gas suspends the plunger in a surface lubricator assembly, preventing it from falling down the production tubing (fig. 62). A surface controller (based on time, pressure, or flow rate) actuates a motor valve in the surface flow line to close and shut in the well. The time that the flow line is open and the well is flowing is referred to as ON time. The time that the well is shut in is called OFF time.

With no gas flow to hold the plunger in the lubricator, the plunger falls freely through the standing fluids in the tubing, typically at 300 feet (91 metres) per minute through gas and 150 feet (46 metres) per minute through liquid. A downhole bumper spring assembly absorbs the impact of the falling plunger. A standing valve is sometimes included in the bumper spring assembly to keep accumulated liquids from exiting the tubing during the shut-in cycle for applications with very low pressures, low liquid volumes, or where fluids tend to fall rapidly out of the tubing string to be reabsorbed into the near wellbore.

When the well is shut in, the pressure in the tubing and casing rises to approach the *static pressure* in the formation. The controller then opens the surface motor valve, releasing the back-pressure above the plunger into the surface flow line. The accumulated pressure in the annulus and wellbore below the plunger drives the plunger and liquid slug up the tubing at high velocity—between 800 and 1,000 feet (244 and 305 metres) per minute. At the surface, the fluid continues into the flow lines while the plunger travels into the lubricator. A spring in the lubricator absorbs the impact of the plunger; the plunger is again held in place by the gas flow until the next shut-in cycle.

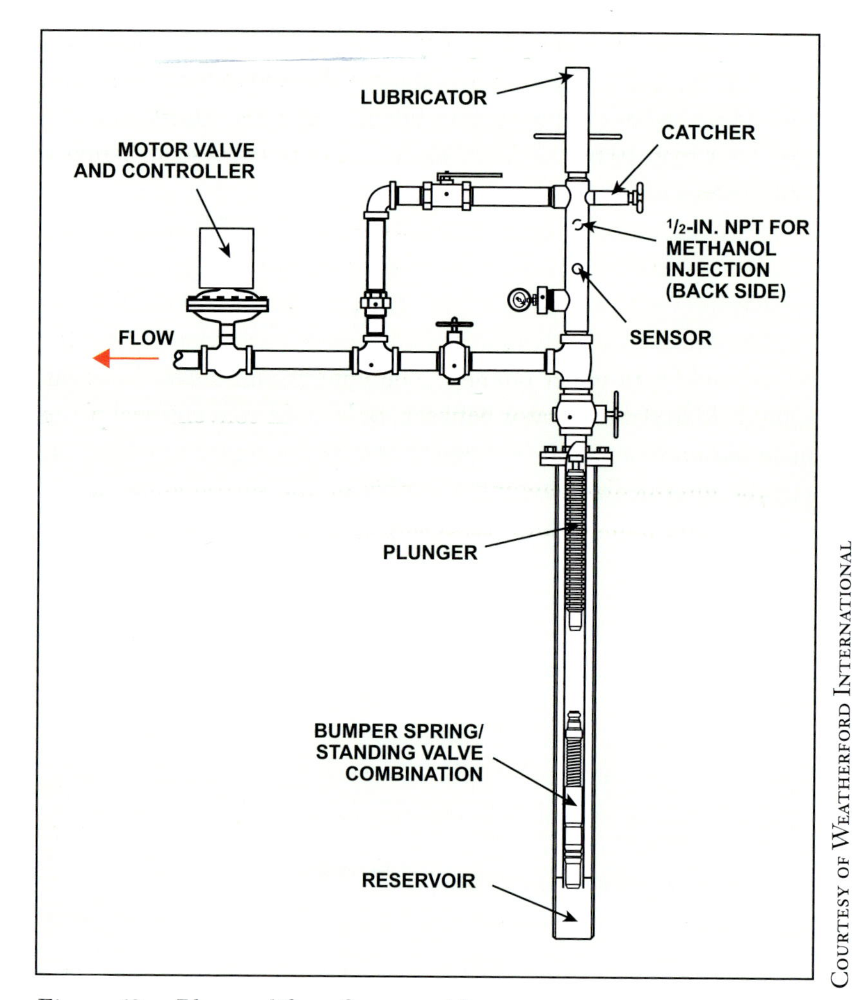

Figure 62. Plunger lift surface assembly

Some lubricators have an automatic catch mechanism that positively secures the plunger in the lubricator between cycles. At the beginning of the shut-in cycle, a motor valve opens the catch mechanism to release the plunger. Upon returning to the lubricator, the catch mechanism again holds the plunger in place until the next shut-in cycle.

The frequency of the unloading cycles and the duration of the shut-in time are dependent upon the application, depth of the bottomhole assembly, and gas-to-liquid ratio of the well. For dewatering gas wells, production is maximized by more frequent shut-in cycles lifting small slugs in order to maintain the lowest possible 24-hour-average bottomhole pressure. Similarly, the goal of oilwells is to maximize the number of cycles to continually lift fluid. In oilwells, it is common to have little or no time between cycles.

Continuous-flow plunger systems are designed to fall while the well is flowing. They are most commonly used at the first stages of *liquid loading* in the production tubing when gas velocities are high (greater than 10 feet or 3 metres per second). In these high-flow applications, it is possible to lift with little or no shut-in time.

Progressive plunger lift (or staged plunger lift) consists of two or more plungers acting in series in the same tubing (fig. 63). The plungers are separated by an intermediate landing assembly, which holds the fluids delivered from the lower plunger until the fluids are retrieved by an upper plunger. The uppermost plunger is captured and held in the lubricator between cycles as in conventional plunger lift. Typically, the lower plungers have open tolerances to hold them at the intermediate landing assembly by the surrounding gas flow. Progressive plunger lift is especially applicable to low-GLR wells in which the hydrostatic head must be minimized and the cycle frequency is expected to be high.

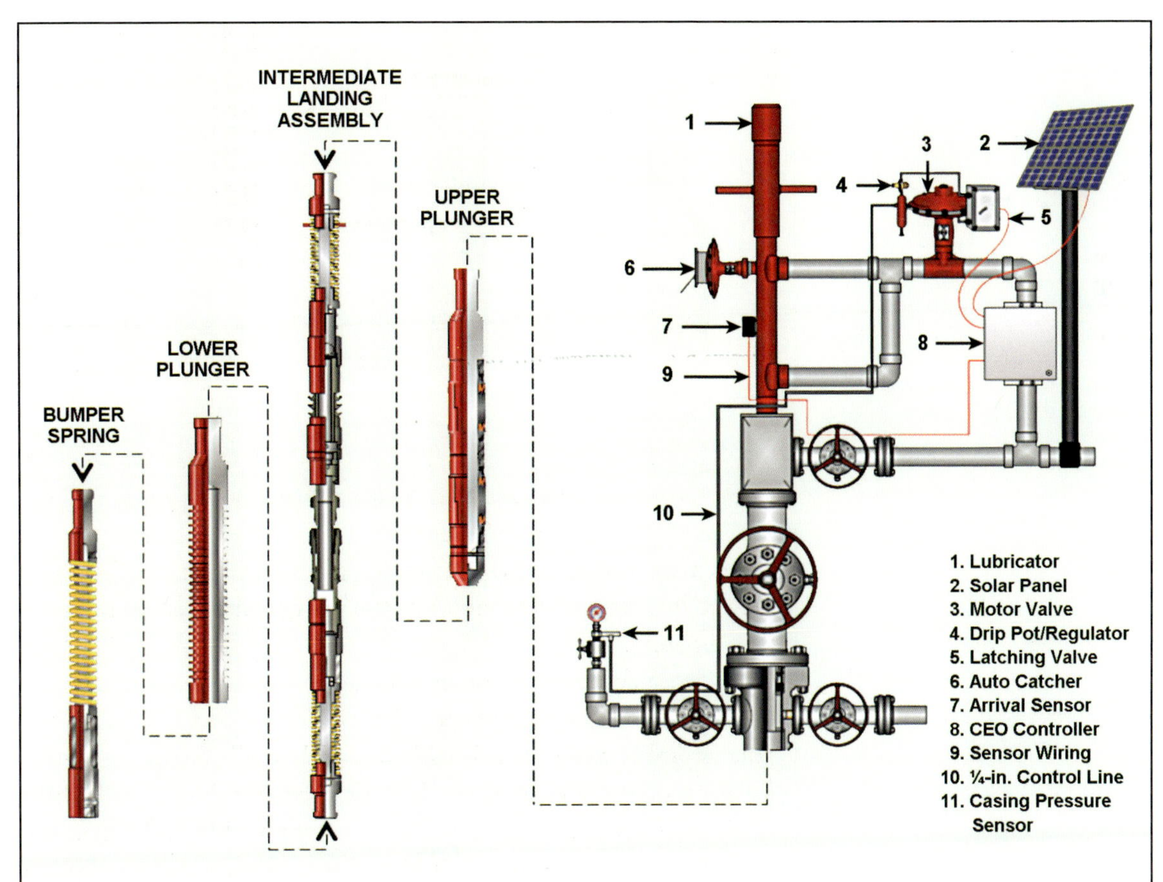

Courtesy of Weatherford International

Figure 63. Progressive (staged) plunger lift

System Components

Pad plungers are very efficient, but they should not be used in wells where particulate matter might foul the pad linkage.

Conventional plungers have no bypass feature and require the well to be shut-in in order for the plunger to fall (fig. 64). Solid ring plungers (sometimes called spiral plungers) have no moving parts or seal elements, so they are effective in wells with abrasive fluids. They rely on a hydrodynamic seal between the plunger and the tubing. The resulting high-fluid fallback results in relatively low efficiency compared to other plunger styles. Brush plungers are effective at keeping the inside of the tubing clean. They are initially efficient, but over time the brushes wear and efficiency drops. Pad plungers have metal pads that collapse when the plunger falls but which actuate to run on the tubing wall during the lifting part of the lift cycle. Therefore, they are very efficient, but they should not be used in wells where

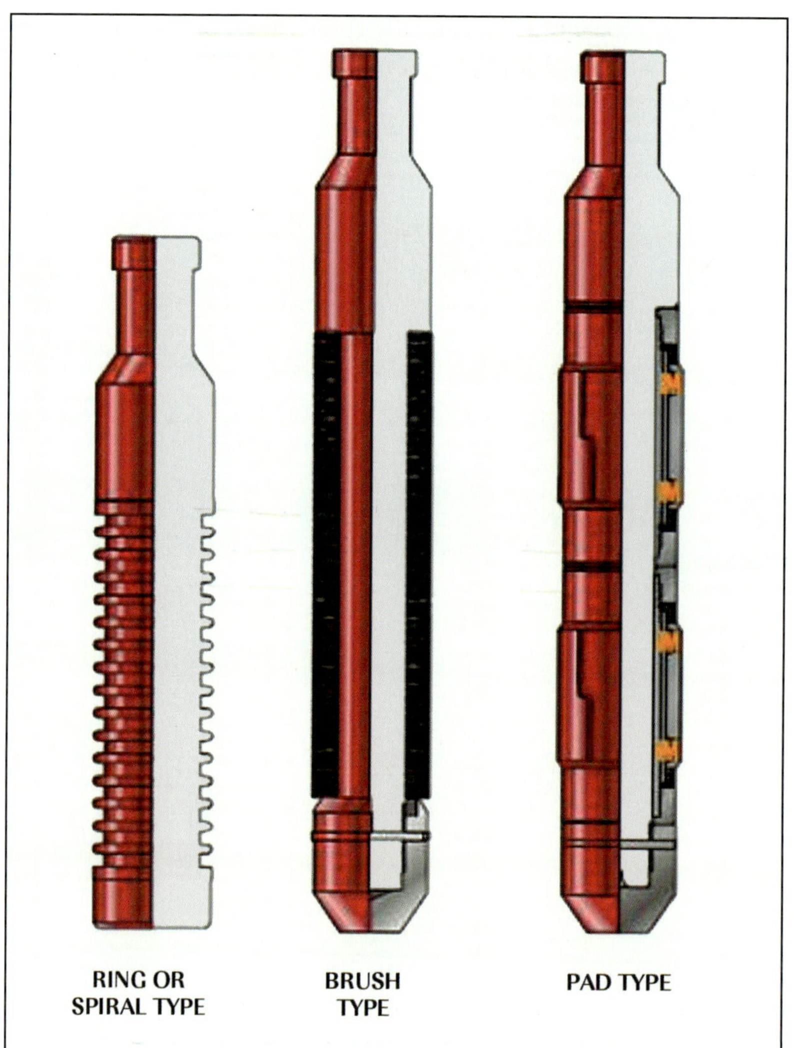

Figure 64. Conventional plungers

particulate matter in the well might foul the pad linkage. Continuous flow plungers have features to internally bypass fluids to increase the plunger fall rate (fig. 65). They can achieve fall rates approaching 1,700 feet (518 metres) per minute in gas and 1,000 feet (305 metres) per minute through liquid.

COURTESY OF WEATHERFORD INTERNATIONAL

Figure 65. Continuous flow plungers

The subsurface assembly includes a bumper spring to absorb the impact of the falling plunger (fig. 66). For wells with low fluid volumes and wells where the fluids fall rapidly from the tubing and into the formation, a standing check valve is included in the subsurface assembly. This keeps liquid on the subsurface assembly and prevents the plunger from running dry.

The surface lubricator includes a spring to absorb the impact of the returning plunger (fig. 67). A sensor in the lubricator sends a signal to the controller when the plunger arrives. Each lubricator also has a manual latch to lock the plunger in place within the lubricator. Lubricators with automatic catch features will hold and release plungers in response to signals from the plunger-lift controller.

The frequency and duration of the cycles are controlled by the plunger-lift controller via actuation of the motor valves (fig. 68).

Courtesy of Weatherford International

Figure 66. Subsurface assembly

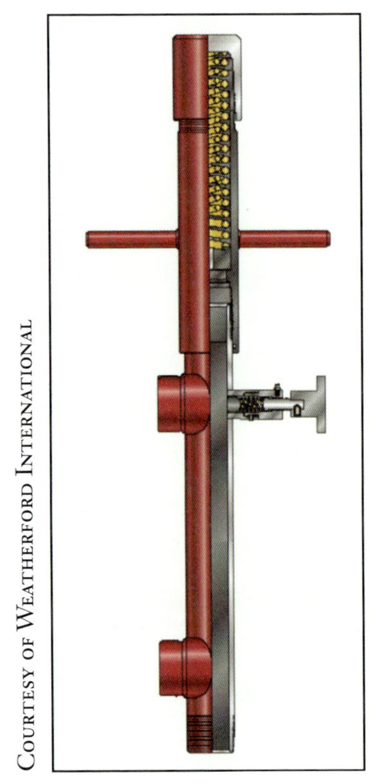

Courtesy of Weatherford International

Figure 67. Plunger-lift lubricator

Courtesy of Weatherford International

Figure 68. Plunger-lift controller

The cycle duration can be controlled based on the interval ON time, interval OFF time, and plunger arrival time. Cycle times can be based on a variety of methods, including preset times, previous cycle arrival times, pre-established well pressure conditions, flow rates, and differential flow conditions, to name a few.

System Design

Liquid loading can be revealed by comparing the actual decline curve to the original decline curve.

To remove liquids from gas wells, the first step is to verify that the well is loading up with liquids. This can be done by comparing the actual decline curve (produced gas volume over time) to the original decline curve if the well has flowed above critical velocity such that gas-flow rates were sufficient to remove all of the water. If the actual decline curve is below the original decline curve, then the well loads with liquids along with some produced gas. In addition, any erratic declines in the decline curve are usually indicators of liquid loading.

If the well is loading with liquids, the criteria in table 4 can be used to determine whether or not there is sufficient gas for conventional plunger lift. Ideally, the annulus should be available to pressurize through the formation when the flow-line valve is closed. If the annulus is isolated by a packer, then additional formation gas-flow rate is required to lift the plunger.

If sufficient pressure gas flow is available, the plunger type can be selected based on the gas velocity in the tubing:

- Over 15 feet (5 metres) per second, a high-speed continuous-flow plunger can be used.
- Over 10 feet (3 metres) per second, a pad plunger can be used.
- Under 10 feet (3 metres) per second, additional calculations should be used to determine whether or not conventional plunger lift can be used (based on achievable differential pressure buildup in the wellbore).

Table 4

PLUNGER LIFT SCREENING CRITERIA

Flowing Wellhead Pressure	**Low Pressure Wells** $P_{FWH} < 250$ psi (1,600 kPa)	**Higher Pressure Wells** $P_{FWH} > 250$ psi (1,600 kPa)
No packer	Q > 500 scf/bbl/1,000 ft ($0.5\ m_g^3/m_l^3/m_d$)	Q > 1,000 scf/bbl/1,000 ft ($1\ m_g^3/m_l^3/m_d$)
Packer	Q > 1,000 scf/bbl/1,000 ft ($1\ m_g^3/m_l^3/m_d$)	Q > 2,000 scf/bbl/1,000 ft ($2\ m_g^3/m_l^3/m_d$)

Summary

Plunger lift takes advantage of produced gas to drive a plunger up the tubing string, pushing a slug of liquid to the surface as it does so. The fluid is lifted using the energy of the formation rather than requiring pressurized injection gas energy from the surface. The result is the lift technology with the lowest cost per volume lifted for low-volume applications. It is most commonly applied to wells producing 1 to 50 barrels per day. Plunger lift is sometimes used in conjunction with intermittent gas lift. During the operation cycle of a plunger-pump system, the plunger falls to the bottom of the well, causing pressure to accumulate below the plunger. The pressure then drives the plunger and a liquid slug up the tubing. The plunger is held in place at the surface until it is time for it to fall again. The frequency and duration of the operational cycle are dependent on the application, depth of the bottomhole assembly, and gas-to-liquid ratio of the well.

Velocity Strings and Foam Lift

In this chapter:

- Typical applications of foam-lift technology
- Operating principles and the role of critical velocity
- Using surfactants to lower surface tension
- How system components work to extend well life

In a flowing gas well, liquids entrain in the gas and accumulate at the bottom of the well. This increases the bottomhole pressure (BHP) in the well and inhibits gas inflow. Also, accumulated liquids displace gas in the near-wellbore formation, reducing gas permeability and hindering gas migration to the wellbore. If flow velocities are sufficiently high, the flowing gas will continuously blow liquids out of the well to keep the well unloaded, or clear of liquids. However, at lower gas velocities liquids accumulate in the wellbore, slowing gas inflow. Eventually the entrained and accumulating liquids can increase BHP to the point that gas production ceases.

Foam lift is primarily a dewatering technology, although effective surfactants have been developed for hydrocarbons.

The term liquid loading refers to the accumulation of liquids in a wellbore that inhibits gas inflow. One way to prevent or relieve liquid loading is to enhance gas velocities; another way is to cause the liquid to foam so that it can be more easily displaced.

Flow velocities can be increased by reducing the cross-sectional flow area of the gas stream. This can be accomplished by flowing the well fluids through reduced diameter tubing (velocity string) or through the annulus around an inserted *dead string* of tubing. Gas flow velocities can also be increased by injecting gas to comingle with produced gas.

Surfactants such as soap or detergent can be used to decrease the surface tension of the liquid so that the liquid foams in the presence of gas and turbulence. The lighter foam is more easily carried to surface by the gas flow.

Typical Applications

Velocity strings and foam lift are often used to extend the life of aging gas wells. Velocity strings can also be effective for enhancing flow velocity in long perforated intervals where inflow velocities are low. If compression is available in the field, gas can be injected to keep gas flow velocities elevated within the wellbore. As well pressures and flow velocities continue to decline over time, velocity strings will eventually become ineffective. Therefore, velocity strings are most effective in applications with gradual production decline rates.

Foam lift, by itself or in conjunction with velocity strings, can be effective for fluid rates as high as 500 barrels per day, although most applications are designed for significantly lower volumes. Foam lift is most typically used for dewatering gas wells, but surfactants have been developed to effectively foam hydrocarbons as well.

Velocity strings and foam have been successfully used to sweep liquids through the lateral sections of horizontal wells. In contrast, subsurface pumps are less effective in lateral sections due to reliability issues and difficulty with gas ingestion. Subsurface pumps are also limited to extracting liquid from one low spot whereas surfactants can deliquify long laterals that have multiple relative low spots.

Operating Principles

The critical velocity can be determined for any well depth using modern computer programs that utilize hydraulic flow correlations.

In a vertical gas flow, critical velocity is defined as the flow velocity at which the flowing drag forces tending to lift a droplet of liquid are balanced with the gravitational force on that droplet. At velocities above this critical velocity, the gas flow would carry the droplet with it. At velocities below this critical velocity, the droplet would fall back. In many gas wells, the gas velocity is below critical velocity, so the wells eventually load up with liquids. Velocity strings are sized to keep gas flow velocities above critical velocity so that the liquids are produced to surface with the gas flow.

Critical velocity is related to the surface tension of the lifted liquids. Reducing the surface tension of a liquid allows the liquid to be displaced by much lower gas flow velocities. Surfactants lower the surface tension of liquids to the point that they foam in the presence of gas turbulence. The foamed liquids become suspended in a bubble film spread out over more surface area and dispersed more evenly within the gas stream. Therefore, the liquid becomes exposed to more of the gas flow energy and is more easily displaced by the gas flow.

Surfactants are designed to initially create foam in the wellbore and then break, or separate, out of the produced liquids before the fluids arrive at the surface. This prevents the surface fluid processing equipment from having to deal with large quantities of foam. The timing in which the surfactant breaks down is dependent on well fluid chemistries and flow rates. Because of this, well fluids should be tested for salinity, hardness, liquid hydrocarbon content, and bacteria to determine the optimum surfactant chemistry and concentration for the application.

System Components

Capillary tubing strings:
- Precisely meter surfactants
- Accurately deliver surfactants to where they can be most effective
- Deliver chemicals that control paraffin, scale, and corrosion

Foam lift can be accomplished using specially formulated soap sticks that are dropped into the well periodically, either manually or by timed release mechanisms. Batch processing is also used, whereby a surfactant is periodically pumped into the well. Results from soap sticks and batch processing, however, can be sporadic. The surfactant concentration varies over time, and the surfactant does not always migrate to where it can be most effective. The surfactant can be continuously injected downward into the annulus if there is no packer in the well, but the treatment is only effective at the bottom of the tubing where there might not be sufficient turbulence for effective mixing.

Capillary tubing strings can be used to improve the effectiveness of the surfactant by continuously metering exact quantities of it and delivering it to where it will be the most effective. To achieve optimum mixing while impacting all the liquid that might load the well, capillary strings can inject surfactant at the producing interval and at the location of highest turbulence. Capillary systems can also be used to deliver other chemicals for paraffin control, scale inhibition, and corrosion control. Chemical surfactants represent the primary cost associated with foam lift systems.

The equations used to define critical velocity were initially derived for surface conditions, but now with modern computer programs utilizing hydraulic flow correlations, the critical velocity determination can be made for any depth in the well. (See figure 69 to learn about critical gas rate related to equations.) With gas production rates known and critical velocity calculated, velocity strings can be sized to achieve gas flow velocities above critical velocity to produce liquids to the surface.

Gas injection can supplement produced gas production to provide sufficient gas flow velocity to remove produced liquids. However, injected gas will increase BHP, which will thereby inhibit gas inflow. Therefore, gas injection should be limited to the amount necessary to remove liquids.

Foam lift can be used to deliquify a well without increasing BHP. Although exact volumes, chemistries, and concentrations vary by application, one gallon of surfactant is generally required for every 20 to 40 barrels of liquid that will be lifted. Water and liquid hydrocarbons react differently to surfactants. Liquid hydrocarbon foams are less stable than water-based foams. Continuous agitation or turbulence is helpful to maintain foam stability of hydrocarbon foams.

Capillary systems consist of a surface chemical tank, metering pump, injection manifold, capillary tubing hanger for the wellhead, ¼- or ⅜-inch (6- or 10-millimetre) capillary tubing, and subsurface chemical injection valve. The capillary tubing can be snubbed into the well under live flowing conditions without damaging it.

Capillary tubing material selection is based on well fluid chemistries. In order of increasing cost, the following capillary tubing metallurgies are currently available: 316L austenitic stainless steel, 2205 duplex stainless steel, 2507 super duplex stainless steel, Inconel 625, and Hastelloy C-276.

Standard capillary tubing can be used for wells deviated to 70°. More highly deviated wells should use heavy-wall capillary tubing to handle the additional forces during deployment and retrieval.

Capillary tubing strings are typically run in the annulus between the tubing and casing; however, systems are available to run the capillary tubing inside of the production tubing (fig. 70). Wellheads can be retrofitted or modified to seal capillary tubing upon closing of the master valve.

To minimize the amount of surfactant required, consideration should be given to increasing the gas-flow velocity by decreasing the size of the tubing or cross-sectional area of the flow channels.

System Design

Capillary systems include:
- Surface chemical tank
- Metering pump
- Injection manifold
- Hanger for the wellhead
- Tubing
- Subsurface chemical injection valve

The critical gas rate is defined as the minimum gas flow rate that will ensure the continuous removal of liquids from the wellbore. The equation most widely used to estimate critical rate is Turner's equation, derived from the spherical liquid droplet model, assuming a constant turbulent flow regime. A slight variation of this equation was proposed by Coleman. More recently, an enhancement of the model was proposed by Nosseir, who considered the prevailing flow regimes, and by Li who, to obtain a match to the behavior of the wells he studied, considered the shape of entrained droplets to be closer to that of a convex bean than a sphere.

Turner's Equation:

$$v_{gc} = 1.912[\sigma^{1/4}(\rho_l - \rho_g)^{1/4}]/[(\rho_g)^{1/2}]; \text{ .. assumed } C_d = 0.44$$

Coleman's Equation:

$$v_{gc} = 1.593[\sigma^{1/4}(\rho_l - \rho_g)^{1/4}]/[(\rho_g)^{1/2}]; \text{ .. assumed } C_d = 0.44$$

Nosseir's Equation-I (transition flow regime):

$$v_{gc} = 0.5092[\sigma^{0.35}(\rho_l - \rho_g)^{0.21}]/[(\mu_g)^{0.134}(\rho_g)^{0.426}];$$

Nosseir's Equation-II (highly turbulent flow regime):

$$v_{gc} = 1.938[\sigma^{1/4}(\rho_l - \rho_g)^{1/4}]/[(\rho_g)^{1/2}]; \text{ .. assumed } C_d = 0.2$$

Li's Equation:

$$v_{gc} = 0.724[\sigma^{1/4}(\rho_l - \rho_g)^{1/4}]/[(\rho_g)^{1/2}]; \text{ .. assumed } C_d = 1.0$$

Gas density can be related to gas gravity (Dake) by: $\rho_g = 2.699 \times \rho_g \times p/[Tz]$

Finally, the critical flow rate can be determined from critical velocity by the expression:

$$q_c = 3.06 p v_{gc} A/Tz$$

where:

v_{gc} = critical gas velocity, ft/sec.
q_c = critical gas flow rate, MMscf/day
ρ_l = density of liquid, lbm/ft^3
ρ_g = density of gas, lbm/ft^3
Υ_g = gas gravity (air = 1)
σ = surface tension of liquid to gas, dynes/cm
μ_g = viscosity of gas, lbm/ft/sec
C_d = drag coefficient (dimensionless)
p = pressure, psia
A = cross-sectional area of flow, ft^2
T = temperature, °R
z = gas compressibility factor, dimensionless

For more advanced information on these equations and how they are used, refer to the following publications from the Society of Petroleum Engineers (SPE):

Turner, SPE 2198 PA
Coleman, SPE 20280
Nosseir, SPE 37408
Li, SPE 70016

Figure 69. Critical gas rate and related equations

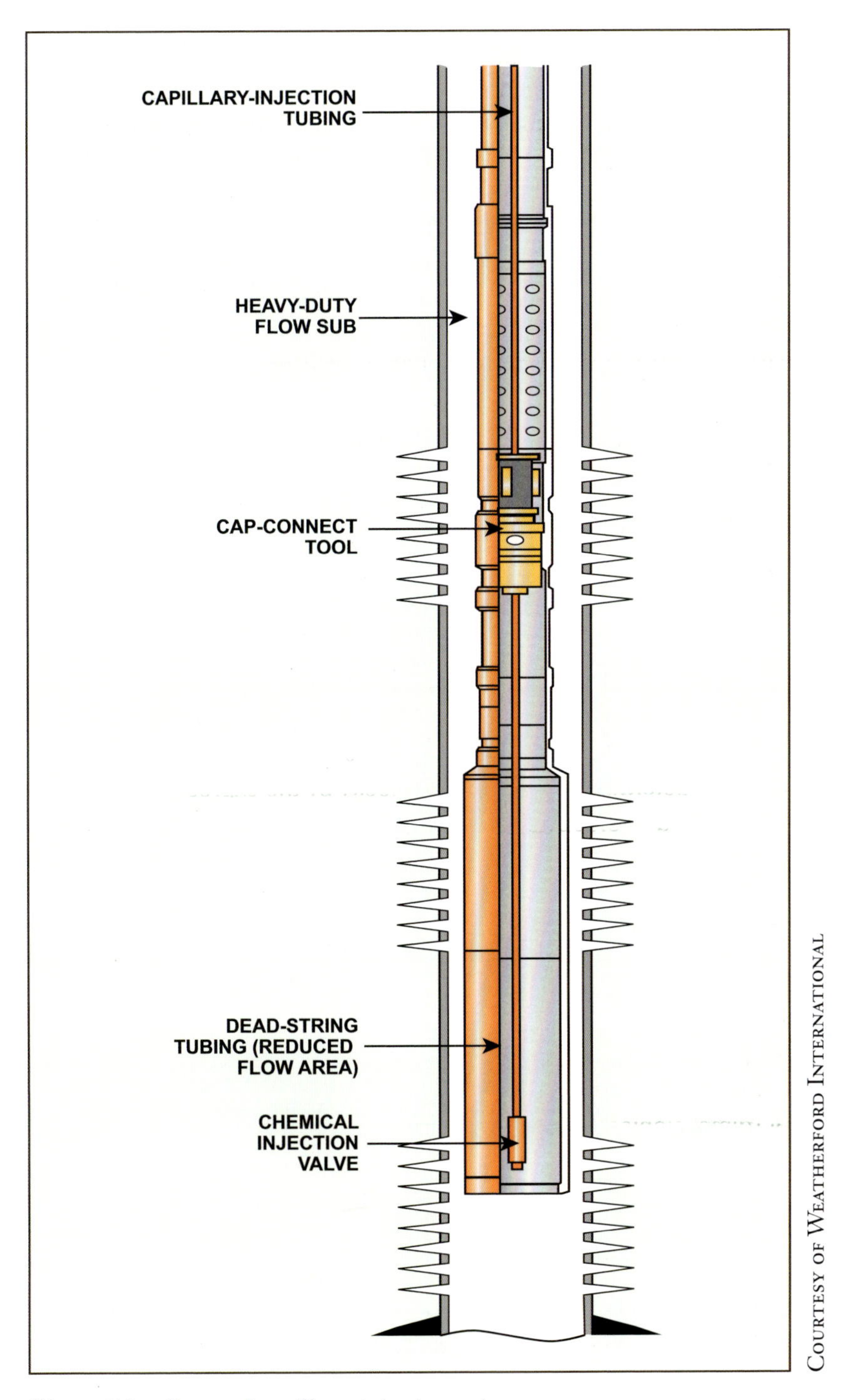

COURTESY OF WEATHERFORD INTERNATIONAL

Figure 70. Internal capillary-injection string

Summary

Velocity enhancement and foam lift, used separately or together, can be effective methods for deliquifying gas wells. Velocity strings reduce cross-sectional flow area, causing flow velocities to increase above critical velocity. Foam lift uses surfactants (soap or detergent) to decrease the surface tension of produced liquids so that they foam in the presence of turbulence. Foam lift can be effective for fluid rates up to 500 barrels per day, although most applications are for significantly lower volumes. Surfactants are introduced into the well in the form of sticks dropped periodically or as a batch pumped periodically. Surfactants can also be pumped continuously through capillary tubing strings.

Hydraulic Lift

In this chapter:

- Typical applications of hydraulic lift
- Configurations of hydraulic-lift systems
- Principles of hydraulic jet and piston pumps
- Surface equipment required for hydraulic lift

In 1932, C.J. Coberly installed the first hydraulic piston pump in Inglewood, California as a solution to pumping oil without using a sucker rod string. Later, Coberly formed Kobe, Inc., and the company was the first to successfully use a hydraulic jet pump to produce an oilwell. Since then, jet pumps have been used to pump up to 35,000 barrels of well fluids per day. Hydraulic pumping represents one of the most flexible forms of artificial lift; it can often successfully produce wells in which other lift technologies have failed (fig. 71).

Hydraulic-pumping systems consist of four basic parts:

- Power-fluid conditioning and supply
- Surface power unit and hydraulic pump
- Piping to transfer the high-pressure power fluid to the subsurface pump
- Subsurface jet pump or piston pump (fig. 72)

The fluid-conditioning system cleans and prepares the power fluid, which is typically a produced well fluid, such as water or oil.

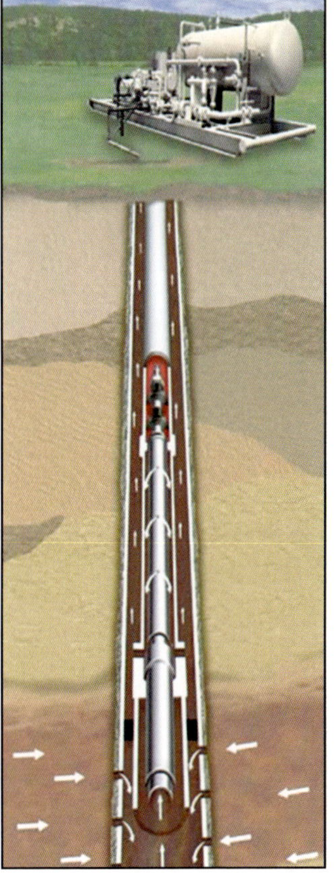

Courtesy of Weatherford International

Figure 71. Hydraulic-lift system

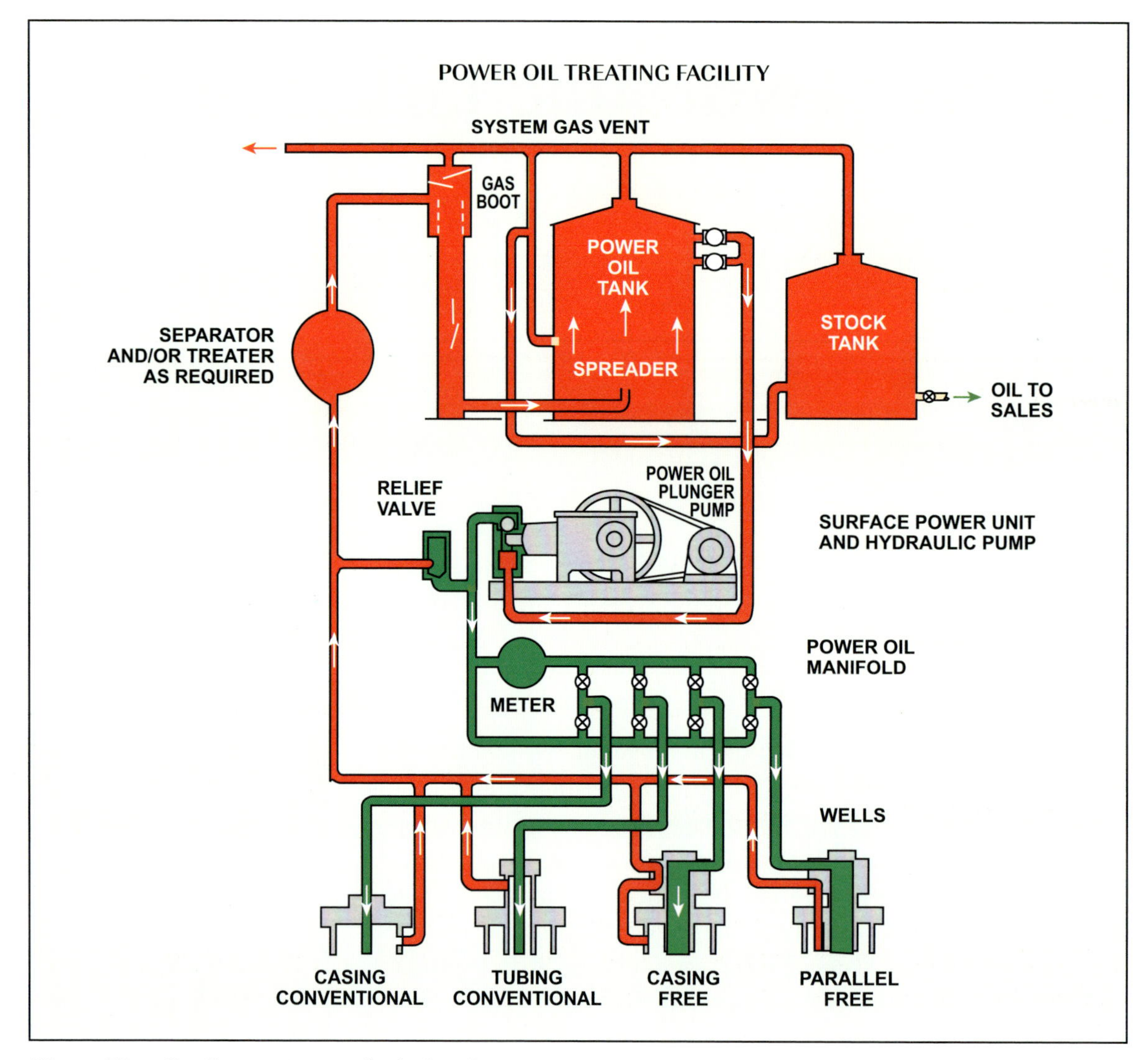

Figure 72. Surface equipment for hydraulic pumping

Hydraulic-lift systems include a subsurface jet pump or hydraulically powered piston pump to generate lift.

The surface power unit and pump pressurize the power fluid. The power fluid is then sent to the subsurface pump through the production tubing, coiled tubing, or through the tubing-casing annulus. The subsurface pump generates lift in proportion to the flow rate and pressure of the power fluid. Jet pumps operate on a venturi nozzle principle with no moving parts (fig. 73). Piston pumps are functionally similar to sucker rod pumps with the addition of a reciprocating hydraulic engine section to stroke the pump piston (fig. 74). The power fluid is typically comingled (open loop) into the produced fluid flow stream returning to the surface. At the surface, the produced fluids can be separated to collect, clean, and re-use the power fluid.

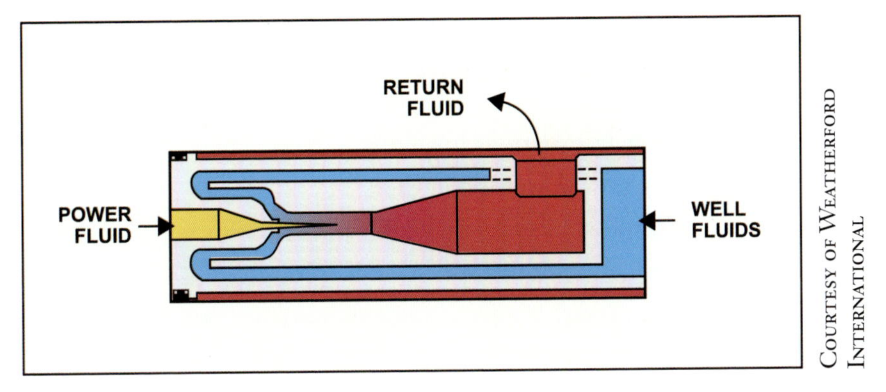

Figure 73. Hydraulic jet pump

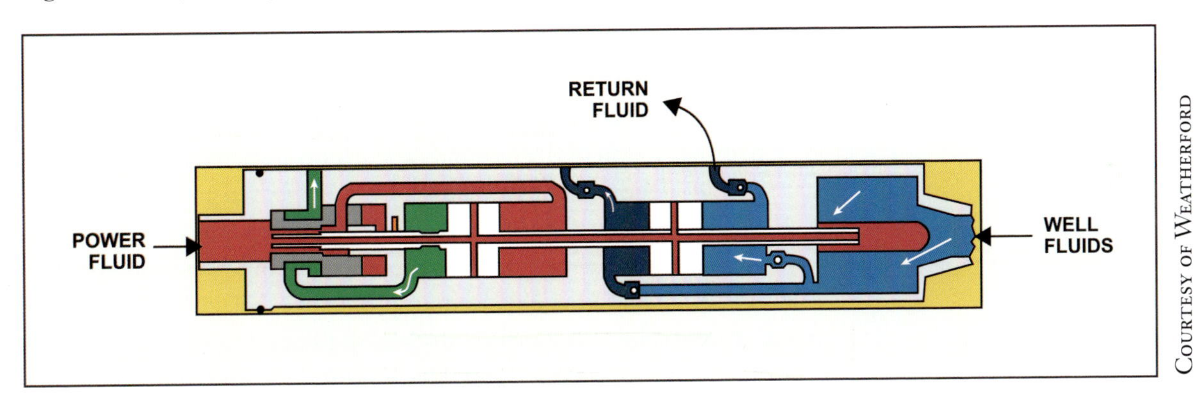

Figure 74. Hydraulic piston pump operation

Typical Applications

Hydraulic lift is arguably the most versatile lift system, capable of handling a wide-range of applications and conditions.

Hydraulic-lift systems are capable of handling the broadest range of application challenges of any artificial-lift technology. They can be configured to be tolerant of gas, sand, corrosive fluids, and a wide range of fluid viscosities. Hydraulic pumps can be used as deep as 17,000 feet (5,182 metres) and can deliver volumes over 35,000 barrels per day, although not concurrently. Insert-style pumps are easily run and retrieved hydraulically, by wireline, or by coiled tubing without any type of service rig. Thus, intervention costs are negligible compared to other lift types. Because there are no rods, hydraulic-pumping systems do not interfere with subsurface safety valves nor do they cause any damage to tubing. They are not affected by crooked and deviated wells. Heated power fluids can be used to help reduce the viscosity of heavy oil.

For all of their advantages, hydraulic-lift systems are used less often than most other lift technologies because of the required surface infrastructure. Surface systems are required to deliver high-pressure power fluids. Hydraulic piston pumps require very clean, filtered fluid, so additional surface fluid processing equipment might be required.

Hydraulic-lift systems are used less often than other lift technologies due to the high systems cost and power requirements.

Centralized systems that provide power fluid to multiple wells can reduce the per-well capital cost. Hydraulic-jet systems require a substantial amount of power. Therefore, hydraulic-jet systems are most applicable in situations in which:

- Low-cost power is plentiful
- Intervention costs are high
- Production downtime must be minimized

System Configurations

Hydraulic pumping installations generally fall into three broad categories according to how the pump is conveyed in and out of the well: free-pump, conventional-tubing-conveyed, and wireline-conveyed. Installations can be further classified by whether the power fluid and produced fluid are comingled (open loop) or kept separate (closed loop) downstream of the pump. Closed-loop systems simplify surface treatment of the power fluid, especially for oil-based power fluids to be re-used with particulate-sensitive hydraulic piston pumps. However, closed-loop systems are more complicated than open-loop systems. With the increased use of jet pump systems that utilize water as a power fluid, closed-loop systems tend to be used less frequently. Accordingly, the following system descriptions are limited to open-loop configurations where the power fluids return to the surface comingled with the produced fluids.

In standard free-pump casing return configurations, the subsurface pump is circulated into or out of the well inside the production tubing (fig. 75). A packer seals the tubing-casing annulus below the pump. Power fluid is injected down the production tubing, and the comingled production and power fluids return up the tubing-casing annulus. The power fluid pressure holds the pump against a bottomhole seating assembly. Reversing the flow so fluid is circulated down the annulus will close a standing valve in the bottomhole assembly. The flow of fluids will then hydraulically lift the pump off the bottomhole assembly and flow it back to surface. Free-pump casing return installations are the simplest and most common hydraulic-pumping system configuration. Since the packer seals the annulus from the face of the formation, all produced gas must pass through the pump. This configuration is typically used in applications with gas-to-liquid ratios of less than 1,000:1; however, production of higher gas-to-liquid ratios is possible.

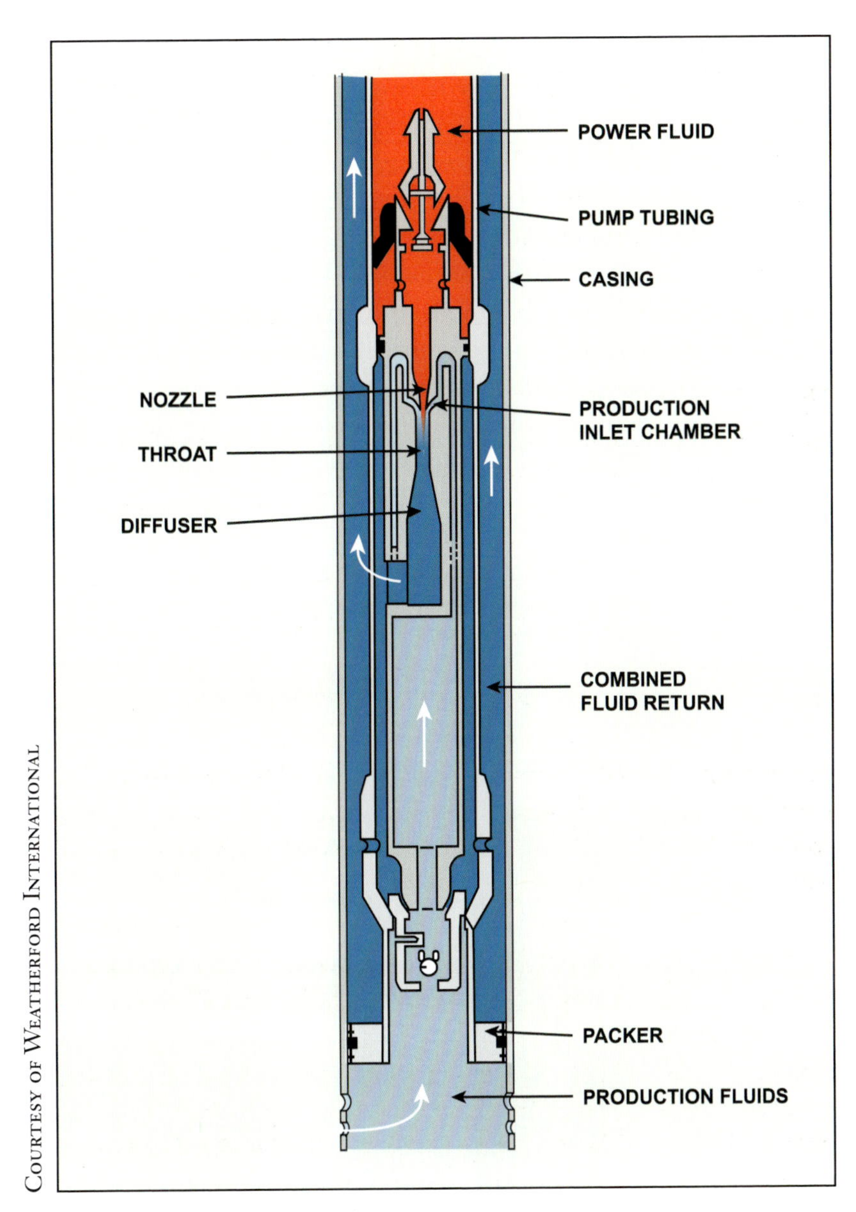

Figure 75. Free pump casing return

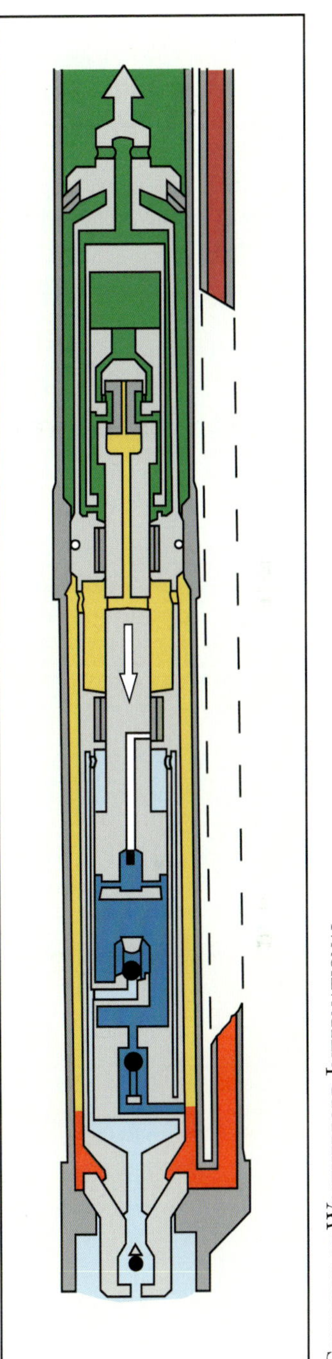

Figure 76. Free pump parallel return

Free-pump parallel return systems use a second string of tubing for the produced fluids (fig. 76). This arrangement includes the basic features of a free-pump system, but it keeps the well liquids off of the casing and allows formation gas to be produced up the casing annulus without going through the pump.

All wireline configurations require the tubing-casing to be isolated from the producing interval by a packer, so any formation gas must be produced through the pump.

Conventional-tubing-conveyed (or fixed-pump) configurations have the subsurface pump run into the well on the lower end of a coiled-tubing string, which is used to conduct power fluid to the subsurface pump (figs. 77 and 78). Fixed-casing conventional configurations return the produced fluids up the tubing-casing annulus. A packer is required to isolate the tubing-casing annulus from the producing interval. Fixed-casing conventional installations are used to accommodate larger production rates in wells with smaller casing. Fixed-insert conventional configurations are similar to fixed-casing conventional configurations, except that a concentric string of production tubing is included between the coiled-tubing string and the casing. In this arrangement, the produced fluids return through the annulus between the coiled tubing and the production tubing. The pump seats against a profile at the bottom of the production tubing so a separate packer is not required. The fixed-insert conventional arrangement keeps the produced fluids off of the casing. It allows formation gas to be produced through the casing annulus, and it can be used in open-hole applications. This configuration is well-suited for dewatering gas wells.

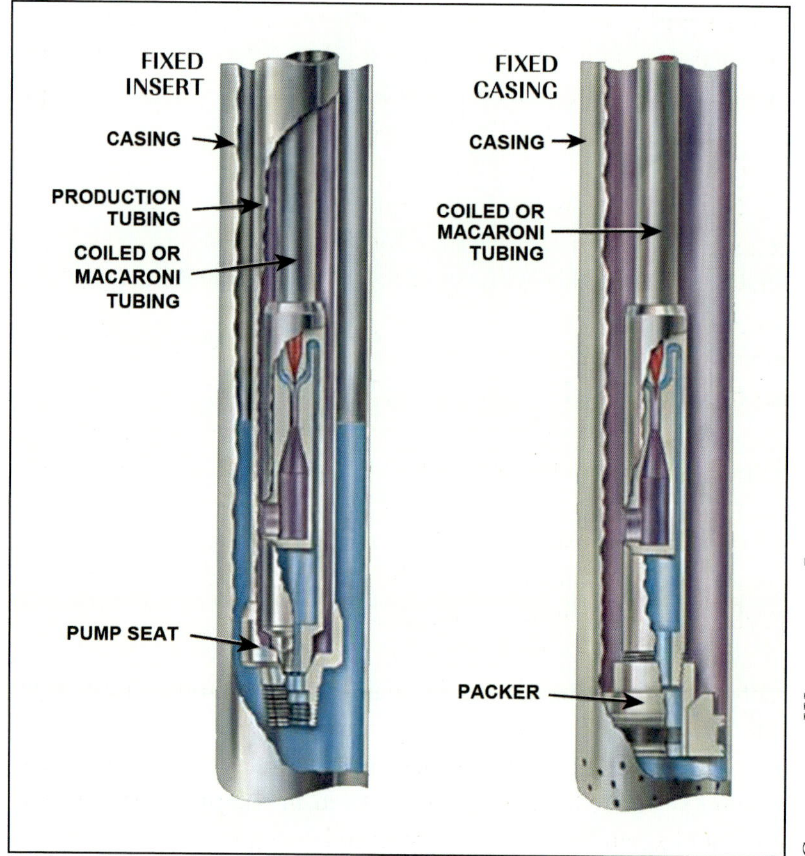

Courtesy of Weatherford International

Figure 77. Tubing-conveyed pumps

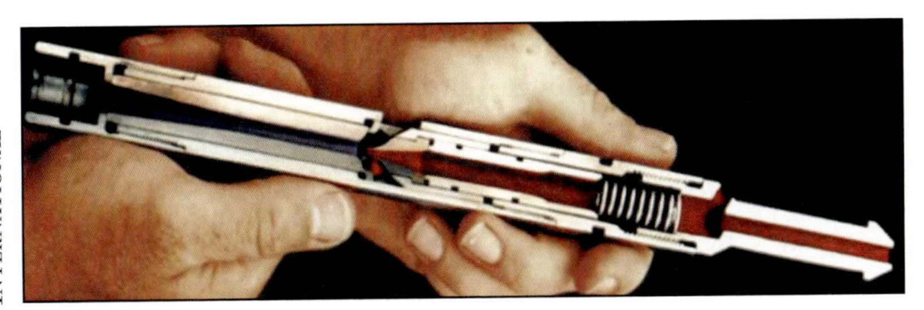

Courtesy of Weatherford International

Figure 78. Coiled tubing jet pump

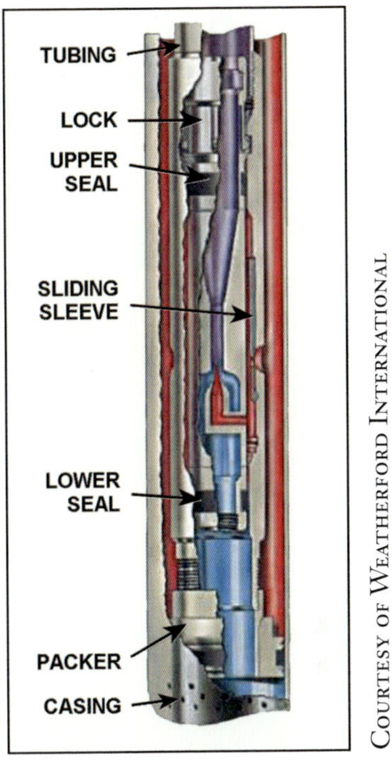

Courtesy of Weatherford International

Figure 79A. Sliding sleeve wireline application

Wireline set configurations are available for installing hydraulic pumps in wells with existing sliding sleeves, existing gas-lift mandrels, and in production tubing without bottomhole seating assemblies. All wireline configurations require the tubing-casing annulus to be isolated from the producing interval by a packer, so any formation gas must be produced through the pump.

For sliding-sleeve wireline installations, the subsurface pump assembly includes upper and lower seals to isolate power fluid, produced fluid, and comingled fluid flow paths (fig. 79A). For standard circulation configurations, the pump is run into the well and is held in place within the sliding sleeve by the power fluid pressure in the production tubing. The comingled fluids are then produced up the casing annulus. The pump can then be retrieved as needed by wireline.

The pump can also be configured for reverse circulation, whereby power fluid is pumped down the casing annulus and the comingled fluids are produced up the production tubing. This reverse circulation arrangement keeps well fluids off of the casing, but casing strength and condition must be considered before pressurizing the casing with the power fluid. A latch feature is required to lock the pump assembly into the sliding sleeve profile, since the power fluid pressure does not act to positively hold the pump in place.

For gas-lift mandrel applications, the subsurface pump assembly includes upper and lower packer elements to seal inside of the tubing above and below the mandrel (fig. 79B). Power fluid reaches the subsurface pump through the production tubing or the annulus. The comingled well fluid and power fluid are produced through the side pocket mandrel port and produced to the surface.

Wireline-type hydraulic pumps can be similarly set in tubing without a sliding sleeve, a gas-lift mandrel, or a bottomhole assembly. The subsurface assembly is configured with hydraulic packer upper and lower sealing elements as per the gas-lift mandrel configuration in figure 79B. A hole can be perforated in the tubing to establish communication between the tubing and casing.

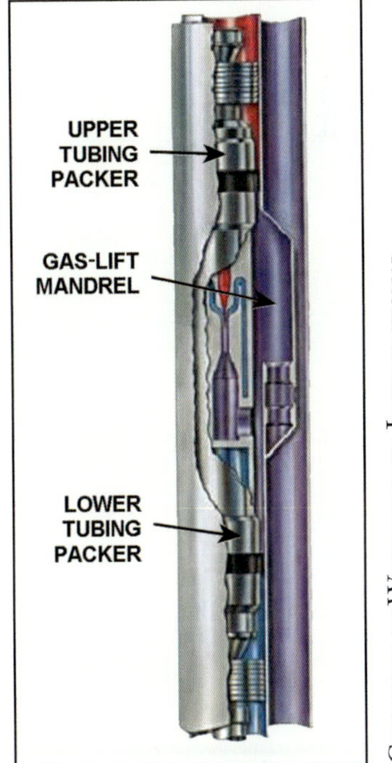

Courtesy of Weatherford International

Figure 79B. Gas-lift mandrel application

Hydraulic Jet Pumps

Jet pumps boost the pressure in the produced fluids by

- Comingling the power fluid and produced fluid
- Transforming kinetic flow energy into pressure, as shown in figure 80 and as described in the following steps:
  1. High-pressure power fluid is injected from the surface at a high rate.
  2. The power fluid passes through a nozzle into the throat of the jet pump. The entrance to the throat is essentially at the same pressure as the production fluid, so the power fluid experiences a large pressure drop as it exits the nozzle. As a consequence of the law of conservation of energy, the pressure drop causes the energy in the power fluid stream to convert to high velocity.
  3. The high-velocity stream of power fluid drags the surrounding produced fluid along with it into the mixing tube.
  4. The comingled power and produced fluids decelerate through the diffuser, causing the flow energy to convert to increased pressure (once again, as a result of the law of conservation of energy) sufficient to boost the fluid to the surface.

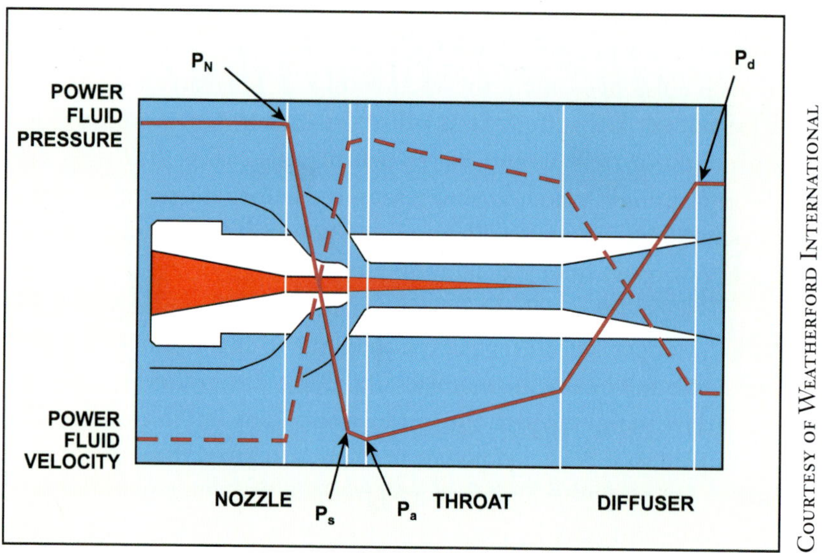

COURTESY OF WEATHERFORD INTERNATIONAL

Figure 80. Schematic of a hydraulic jet pump

The actual time that the fluid is in the pump is very short due to the high velocities. Also, the flow is in a relatively straight line, with very little fluid shear. Therefore, there is little or no emulsification of the production fluids as a result of the jet pump itself, despite the comingling of fluids. Any emulsified produced fluids at the surface would more likely be the result of a surface choke, chemical additives, or acidizing the formation with calcium carbonate ($CaCO_3$) and similar chemicals.

Jet pumps are highly reliable and tolerant of particulate matter, but they consume high energy to accelerate the produced fluids into the flow stream.

The amount of energy that can be transmitted into the production fluid from the power fluid is dependent upon the power fluid mass flow rate. For this reason, gas is not a suitable power fluid for jet pumps because it has low mass. Gas has been used in some dewatering applications using jet pumps, but such installations essentially use jet pumps as exotic gas injection valves to achieve critical lift velocity rather than using the jet pump for mass transfer of energy.

Jet pumps have no moving parts. The only elastomers are O-rings, seal rings, and external swab cups for static sealing. Therefore, they are highly reliable and are very tolerant of particulate matter and elevated fluid temperatures. They are effective across a wide range of produced fluid viscosities and can use heated and treated power fluids to further lower viscosity and dissolve solids.

Jet pumps have lower efficiency compared to other lift technologies primarily due to the energy required to accelerate static produced fluids to high velocity in a very short time span. However, the significant advantages of jet pump systems often justify the higher energy costs necessary to power the pumps.

Hydraulic Piston Pumps

Hydraulic piston pumps are sensitive to scale and particulate matter.

Hydraulic piston pumps include a hydraulic engine-type section to convert the power fluid continuous flow and energy into reciprocating motion (fig. 81). The power fluid causes the engine piston to stroke. At the end of the stroke, the piston shifts the engine valve, which diverts the power fluid to cause the engine piston to stroke in the reverse direction. When the engine piston is fully retracted, the engine valve shifts again to begin the next piston extension stroke cycle.

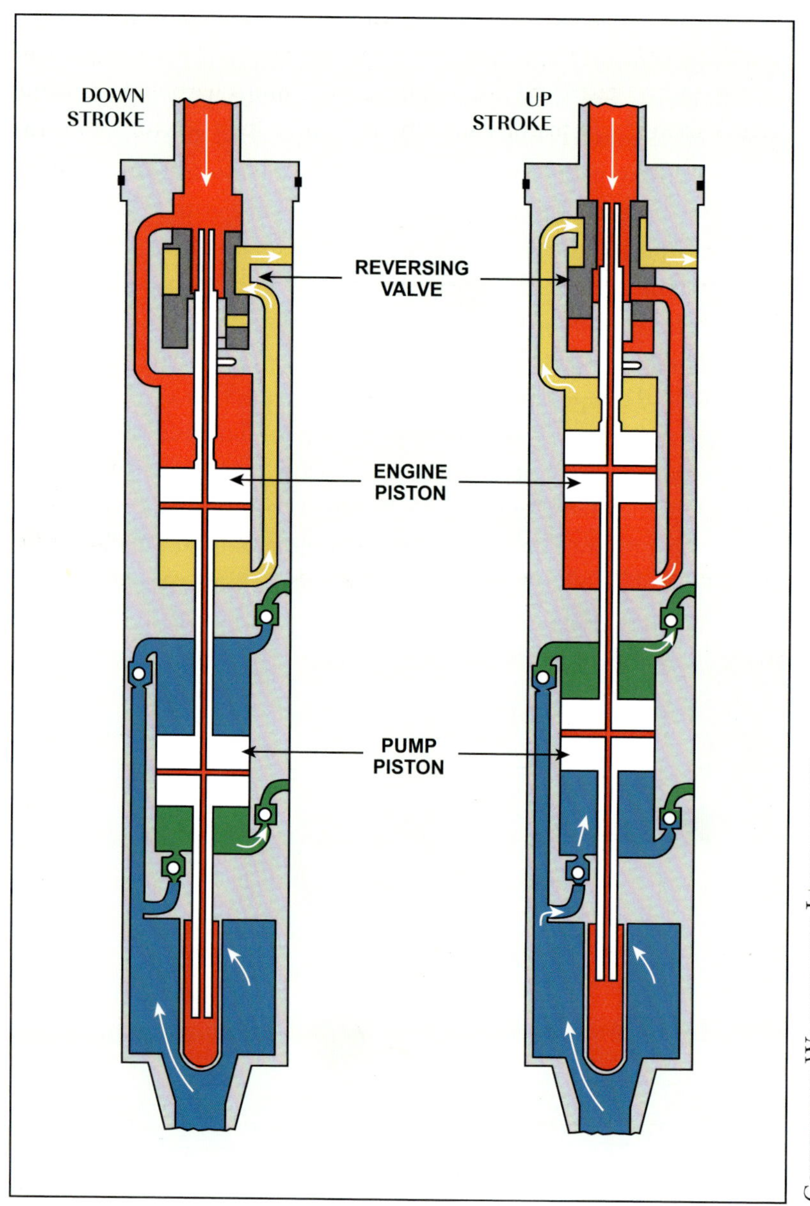

Figure 81. Engine-type section of a hydraulic piston pump

The engine piston is coupled with a piston in the pump section so that both pistons stroke together. As the pump piston strokes, production fluid in the pump chamber becomes pressurized and discharges across a check valve. On the return stroke, production fluid refills the pump chamber through another check valve. The pump section can be single-acting and functionally similar to rod pumps, or it can be dual-acting and pump in both stroke directions.

The speed of the pumping action is in proportion to the flow rate of the power fluid. The pressure boost capacity of the pump is in proportion to the pressure of the power fluid and the relative areas of the engine piston to the pump piston. Pumps with engine pistons larger than pump pistons amplify the power fluid pressure to create higher discharge pressures in the produced fluids, but the discharge volumes are less per stroke. This arrangement would be beneficial in deep applications. When the engine piston is smaller than the pump piston, the power fluid flow rate is amplified to output higher produced fluid discharge flow rates, but the pressure capacity is decreased. This arrangement is beneficial in shallow wells where pressure capacity is less important than production rates.

Hydraulic piston pumps are positive-displacement devices and are therefore more efficient than jet pumps. They are precision devices with many parts built to very close tolerances. Clean power fluid is necessary for piston pumps. They are best applied in applications without significant amounts of scale, particulate matter, or gas.

Surface Equipment

The subsurface pumps are powered by high-pressure fluid that is pumped from the surface. The power fluids are usually produced liquids that have been conditioned. Produced water with surfactant is the most commonly used power fluid, especially for jet pumps. Conditioned oil is often preferred for hydraulic piston pumps in order to take advantage of the oil lubricity. Power fluids can be heated to help reduce the viscosity of produced fluids.

The surface systems condition the power fluid, pressurize the fluids, and control the pumping rate. They can be self-contained skid units that power one subsurface pump, or they can be large central power plants, supplying power fluid to multiple wells (figs. 82 and 83).

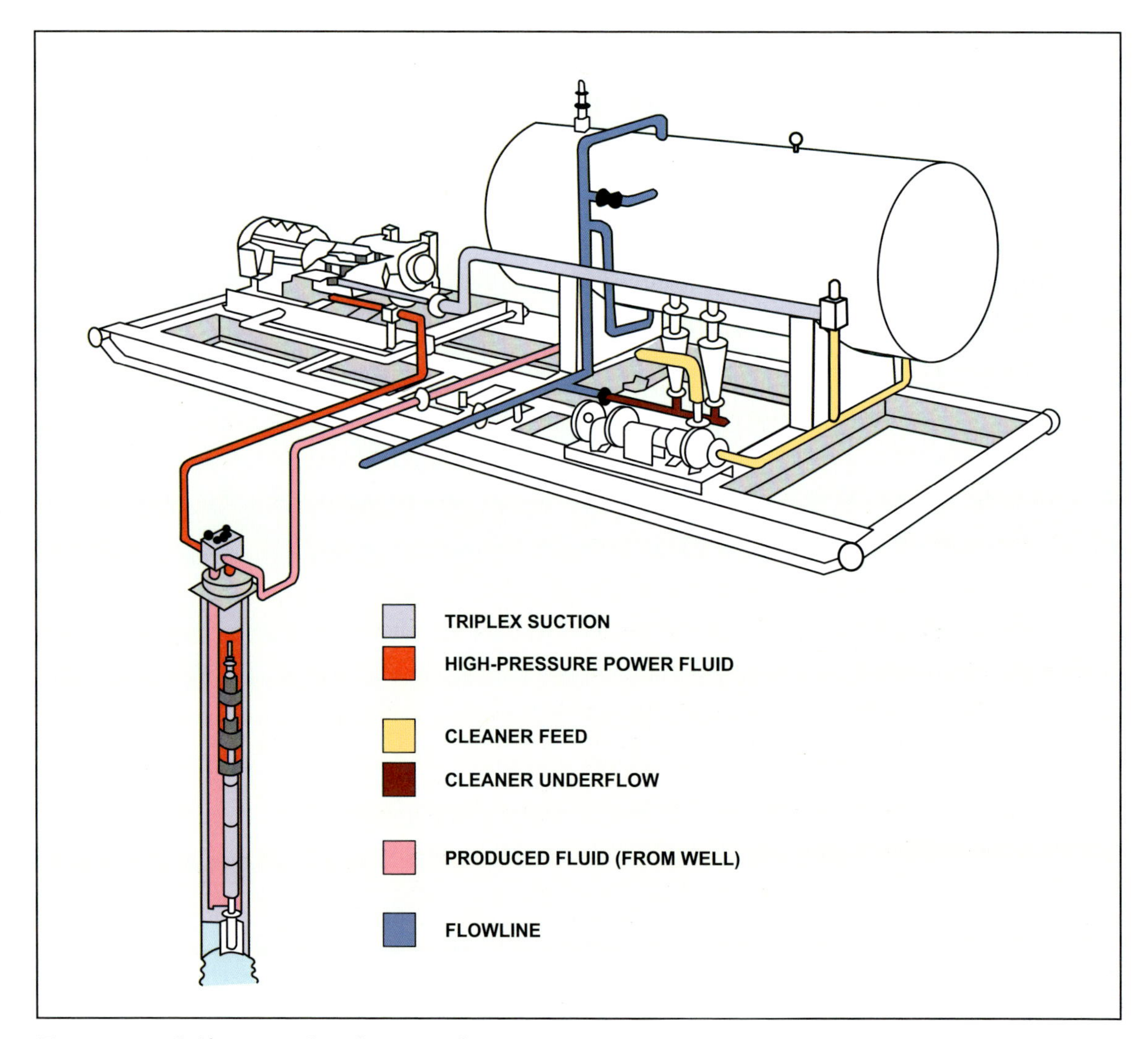

Figure 82. Self-contained surface unit of a hydraulic piston pump

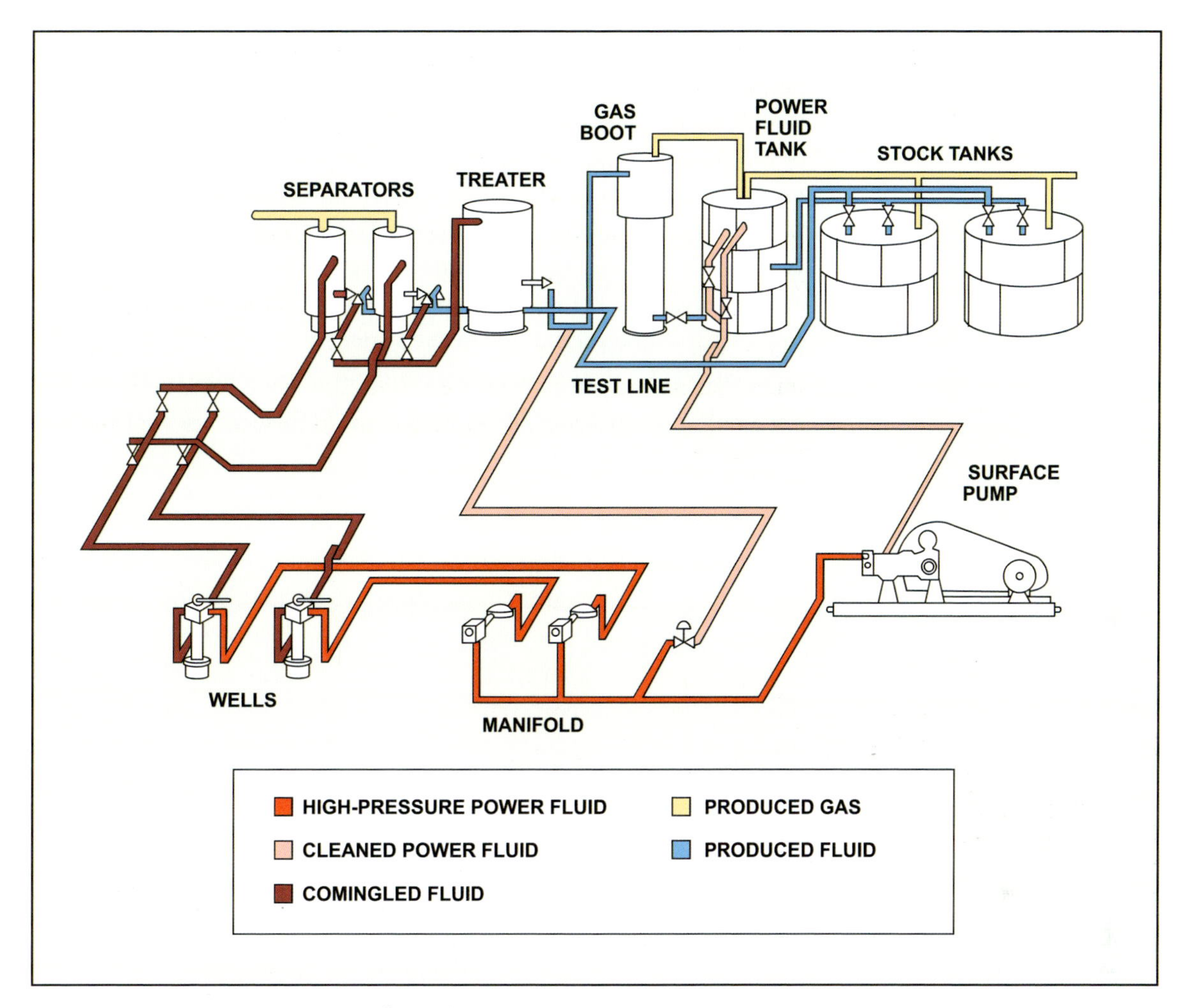

Figure 83. Fluid supply plant of a hydraulic piston pump

Fluid conditioning components can include various types of separators, tanks, heaters, circulating pumps, and associated plumbing. The high-pressure pumps can be multi-cylinder, plunger-type pumps, or they can be ESP-style, horizontal-centrifugal pumps. They can be powered by electric motors, gas, or diesel engines. Wellhead control valves in a control manifold regulate the circulation of power fluid into the well and reverse circulation to return a free-pump to the surface. The surface assembly also catches and holds the returned free-pump and has a bleeder valve to relieve pressure so the pump can be removed.

Summary

Hydraulic-jet pumping systems are capable of handling the broadest range of application challenges of any artificial-lift technology. Hydraulic pumps use high pressure liquid circulated from the surface to power subsurface pumps. Hydraulic jet pumps operate on the venturi nozzle principle. The power fluid passes through a subsurface nozzle, which causes the energy of the power fluid stream to convert to high velocity. Production fluid mixes with the power fluid. The high flow velocity energy is converted to high pressure in the diffuser sufficient to flow the comingled fluids to the surface.

Hydraulic piston pumps are functionally similar to sucker rod pumps, except power fluid is used to transmit power to the pump instead of using a rod string. A reciprocating hydraulic engine-type section is used to convert the power fluid's continuous-flow and energy into reciprocating motion.

The surface equipment of a hydraulic pumping system will include various types of separators, tanks, heaters, circulating pumps, and associated plumbing. The high-pressure pumps can be multi-cylinder plunger type pumps, or they can be ESP-style horizontal centrifugal pumps. They can be powered by electric motors, gas, or diesel engines.

Production Optimization

In this chapter:

- Maximizing production throughout the life of a well
- Addressing factors that can hinder production
- How systems collect and transmit well data
- Key elements of a production-optimization system

The previous sections of this book all deal with selecting, designing, and effectively applying lift technologies based on assumptions about how the reservoir will deliver fluids. In reality, production always varies somewhat from what was expected because well conditions, inflow volumes, and fluid phases change over time.

The artificial-lift system should be adjusted as needed to match the inflow rates from the reservoir. As fluid production declines, the lift system pumping rate must be similarly reduced or the well might become pumped dry of fluids, causing damage to the lift systems and potentially to the reservoir. In other situations, improved reservoir management techniques can increase reservoir deliverability, but production might then be constrained by lift system performance. Until the lift system is adjusted to produce at the new deliverability rates, the lift system can cause an undetected bottleneck. Damage to lift systems and lost potential production from suboptimum lift performance are easily prevented but are often overlooked by manual surveillance practices.

Effective production optimization should:
- Maximize hydrocarbon recovery
- Avoid equipment failures to reduce downtime
- Reduce operating costs
- Ensure health and safety
- Protect the environment

Production optimization refers to managing the production of hydrocarbons over the life of the well as conditions continuously change. More formally, it is managing the decline curve to meet the production goals of the owner of the hydrocarbon asset. Production optimization consists of the following elements:

- Surveillance and monitoring to detect what is happening
- Analysis to determine why it is happening
- Design of solutions to improve production
- Asset management to determine where and when to allocate resources
- Reporting to understand trends

The typical results of effective production optimization include maximizing hydrocarbon recovery; reducing downtime by avoiding equipment failures; reducing operating costs; and improving protection of health, safety, and the environment. For these reasons, it is often more profitable to invest in field-wide production optimization than to simply upgrade lift hardware alone.

Typical Applications

The type of production optimization varies according to the value and complexity of the lift system. Production-optimization systems can be grouped into three main categories (fig. 84).

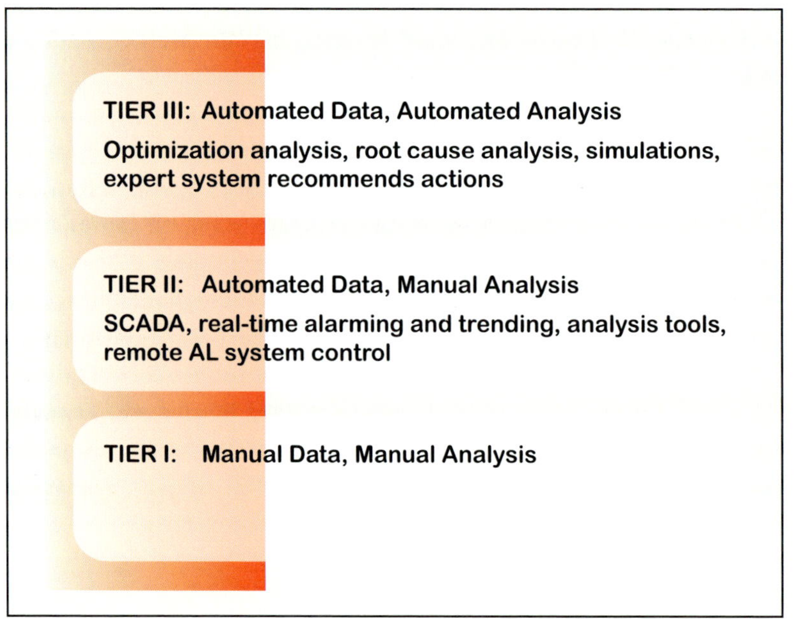

Courtesy of Weatherford International

Figure 84. Main categories of production optimization

Tier I: Manual Data, Manual Analysis

A Tier I production-optimization technique involves any well-status surveillance, monitoring, and subsequent trouble-shooting application that is performed manually, either as part of a drive-by routine or in response to problems. All performance analysis is done manually with manual data input. Lift systems can include controllers that automatically shut down or change the lift system operation in response to problematic conditions such as liquid pump-off or overloads (well-site automation). When well problems are observed, actions are taken to restore proper production. This type of production optimization is reactive in nature but requires the least investment.

Tier III systems should be considered for managing large assets that include many wells.

Tier II: Automated Data, Manual Analysis

In a Tier II scenario, data is automatically collected, organized, and displayed to facilitate analysis by technical personnel who are remote from the well site. Remote surveillance and monitoring, real-time alarming, manually initiated analysis tools, and remote lift system control are all Tier II production-optimization techniques. Alarms can advise field engineers of problematic well conditions before they deteriorate to the point of shutting down production. Personnel can then determine the priority of issues to be addressed. Many Tier II analysis tools are available to the operations engineer. The engineer initiates specific analyses and then determines the best solutions based on the analysis results. Remote system controls allow the operator to adjust lift system performance to maintain optimum production levels. Tier II automation is intended to identify anomalies before they escalate.

Tier III: Management by Exception

Tier III systems go beyond simple display of exception conditions to highlight the most critical issues so that they can be addressed first. In some cases, automatic analysis tools are added to Tier II automated data collection systems to display trends and even suggest causes of exception conditions. Trending can help reveal whether problems are sporadic or chronic and when operator intervention will be required. Although the exception conditions and analyses can be automatically generated by the system, personnel must initiate any changes to system parameters and controls. Tier III systems are critical for managing large assets with many wells.

System Components

Sensors

Sensors provide information on the current status of production and the lift system at specific wells. They measure temperatures, pressures, fluid levels, vibration, flow rates, operating speeds, forces, loads, water cut, and other important conditions. Sensors can be located subsurface to reflect conditions in the well, or they can be at the surface (figs. 85 and 86). Some surface measurements are used to infer conditions in the well.

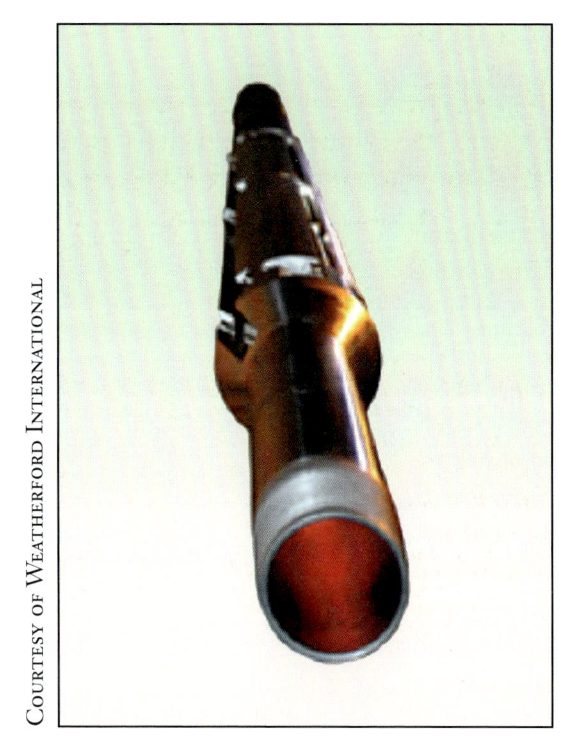

Courtesy of Weatherford International

Figure 85. Subsurface flowmeter

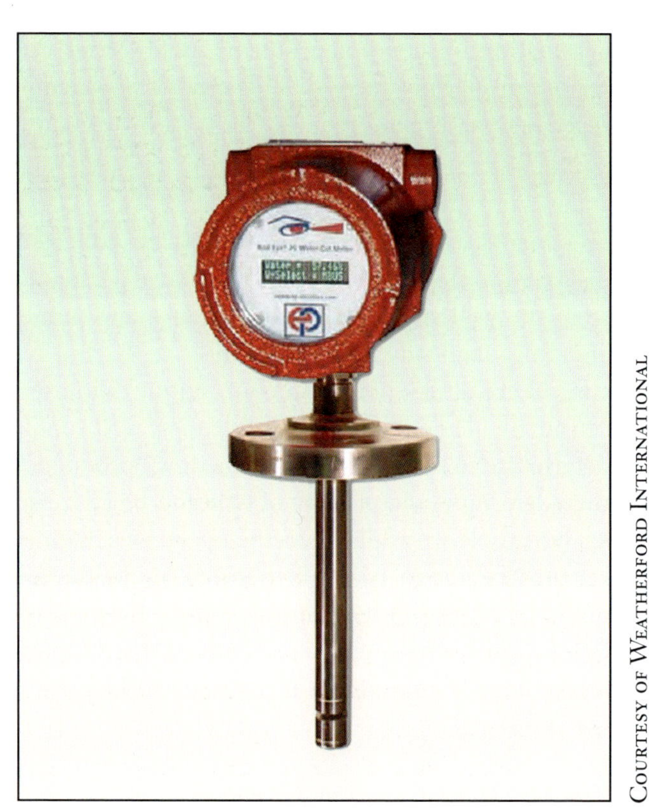

Courtesy of Weatherford International

Figure 86. Watercut meter

Electronic and fiber-optic sensor systems provide data highways within the well for numerous sensors. Electronic subsurface sensors receive power and transmit signals through electric cable that is usually strapped to the tubing. In distributed fiber-optic systems, the fiber cable itself is the sensor and can provide data from any point along the length of the cable (fig. 87). Surface units send optical signals through the fiber cable and analyze the returning or reflected signal to determine pressures, temperatures, vibration, flow rates, and similar well conditions.

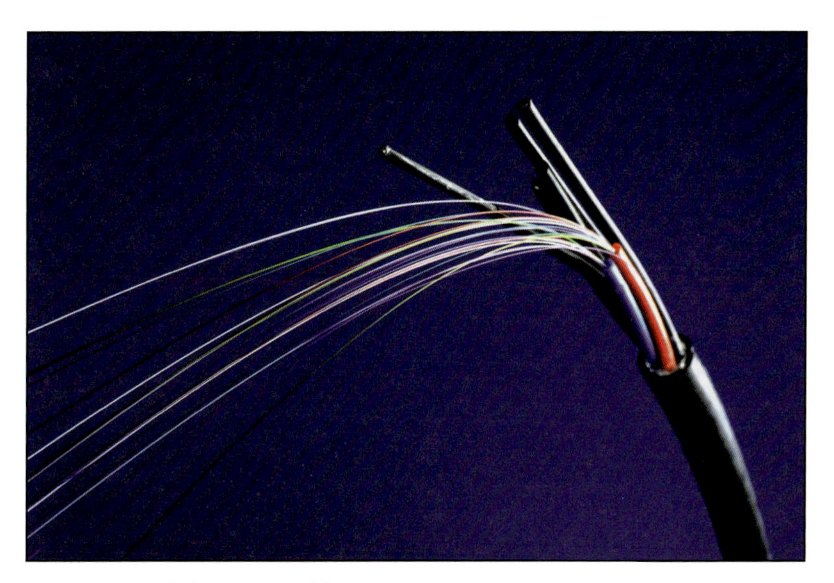

Figure 87. Fiber-optic cable

Well-Site Intelligence

Artificial-lift controllers, intelligent remote terminal units (RTU), and variable speed drives (VSD) interpret and display sensor data at the well site and can be programmed to adjust lift system operations to respond to specific operating conditions. Rod pump controllers can display the rod pump dynamometer cards, which graphically show rod load versus rod position. Controllers can recognize impending pump-off conditions by sensing changes in the rod load that reflect declining well pressures or fluid levels. They can recognize stuck pumps or other potential overload conditions by sensing force, torque, or amperage. The controls can then respond by shutting the system down or adjusting the system speed. The controls can provide a soft start in which systems are brought up to speed in a controlled manner consistent with the system capabilities.

When used with SCADA as part of a Tier II or Tier III system, controllers and VSDs can respond to remote commands to modify lift system operation. Also, field personnel can regularly adjust the control parameters that the controller operates under to assure optimum performance of the well.

SCADA

SCADA systems collect and send data to remote locations using an intricate network of radio, satellite, and computer technologies.

Supervisory Control and Data Acquisition (SCADA) systems collect and communicate sensor data to remote locations (fig. 88). At the well site, a remote terminal unit (RTU) analyzes and stores the signals it receives from the sensors. A remote communications device such as a radio, cell phone, or satellite link periodically sends the data to a remote applications server. The server then distributes the information through a network to operations personnel for analysis. The same SCADA system can be used to send information back to programmable logic controllers (PLCs) and VSDs at the well site, thereby avoiding unnecessary well-site visits.

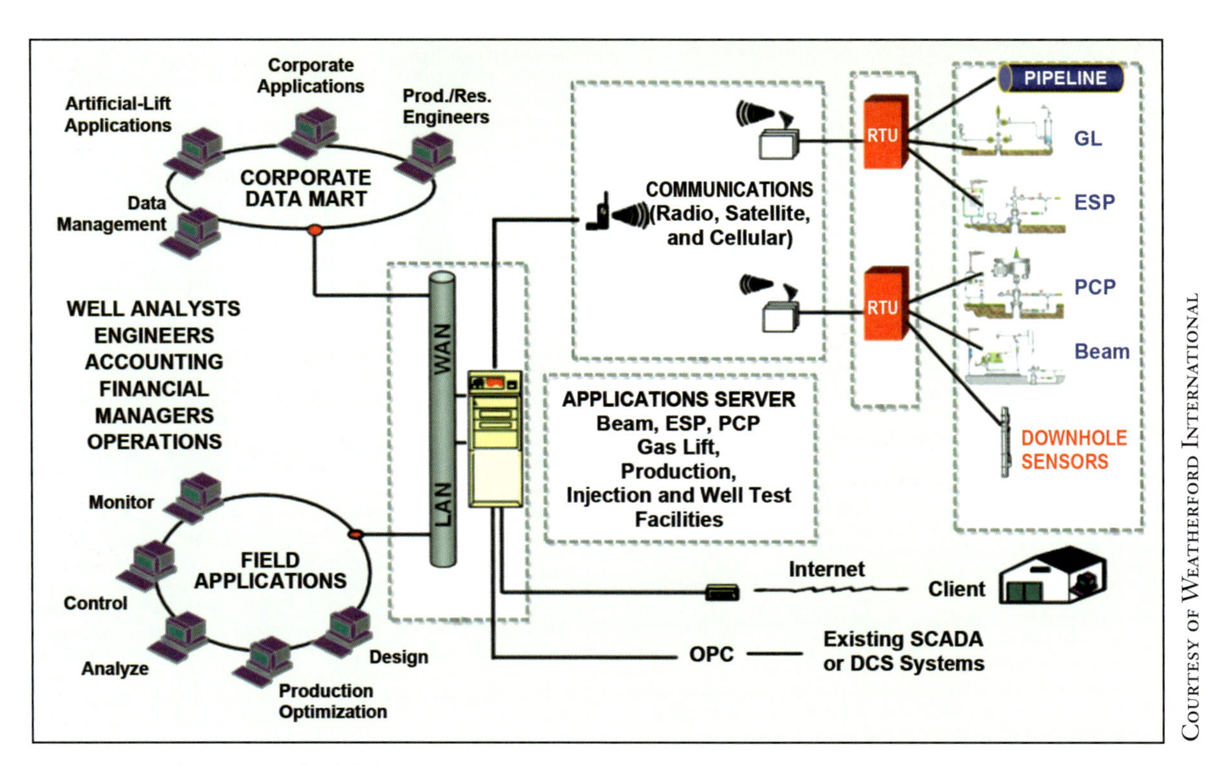

Figure 88. Schematic of a SCADA system

Desktop Intelligence

Desktop intelligence for operations engineers includes remote display and trending of sensor data, alarms, prioritization of status conditions, performance analysis, simulations, expert systems, and sending commands to the PLCs and VSDs to adjust operating rates and parameters (fig. 89). Desktop intelligence can also include monitoring and analysis of production activity for nontechnical personnel such as accounting and corporate groups. The data received can be used in a work management system to provide a head start for repairs and maintenance.

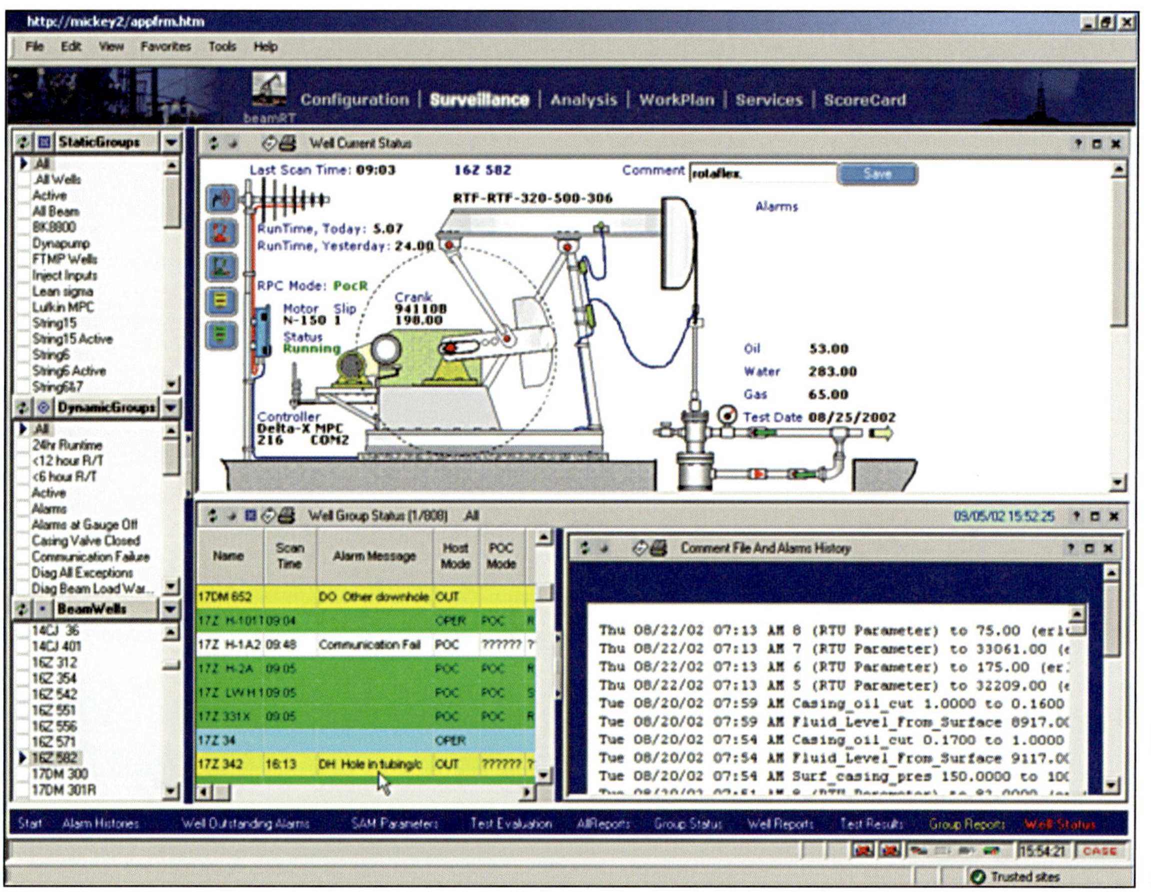

Courtesy of Weatherford International

Figure 89. Example of the user interface for a desktop intelligence software system

Integrated Functionality

Some suppliers offer fully integrated systems that can be implemented as modules to facilitate gradual system implementation (fig. 90). These systems can include modules to model production and resources across an entire field rather than just on a well-by-well basis. Some modules can help operators manage limited resources such as steam, water, or injection gas, or to balance production through limited surface fluid treatment systems. Integrated systems can also provide robust peripheral functionality in areas beyond optimizing production, such as reservoir analysis, production monitoring and forecasting, field inventory management, field service and maintenance planning, and lift system tracking. The single data warehouse structure of the integrated systems reduces time and errors associated with multiple data entry into multiple systems.

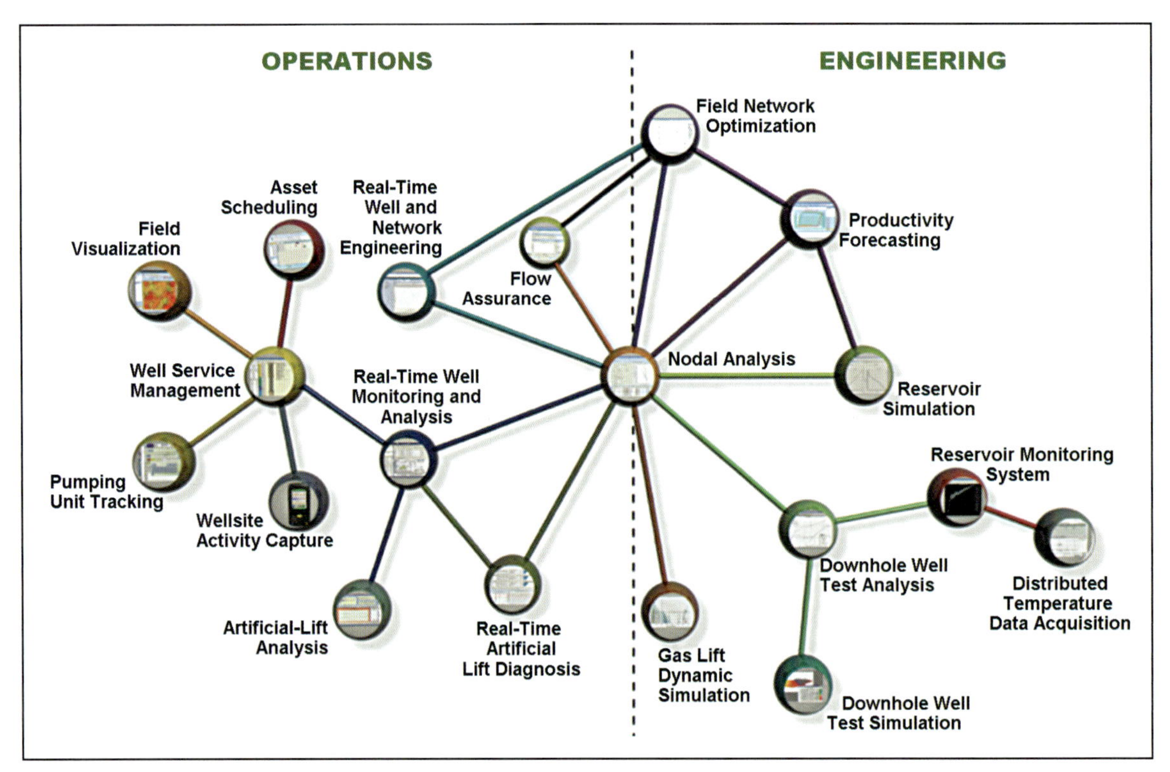

Figure 90. Example schematic of an integrated system

System Design

Most producers begin with Tier I systems and gradually add functionality over time as they recognize the value that these systems provide. However, most of the value of production optimization occurs at the Tier II level by eliminating the operating expenses, improving activity prioritization, and increasing protection of health, safety, and the environment. For this reason, the design of production-optimization systems should be implemented with Tier II or Tier III systems as the ultimate goal.

Many system suppliers provide products that are compatible with other brands of hardware and various software packages. The operator must assure that system components remain compatible as more functionality is added over time.

Tier II and Tier III systems:
- Reduce operating expenses
- Improve activity prioritization
- Improve safety
- Protect health and the environment

Summary

Production optimization is the management of the production of hydrocarbons over the life of the well as conditions continuously change. Production optimization consists of surveillance and monitoring in addition to analysis, solution design, asset management, and reporting. Production-optimization systems can be grouped into three main categories: Tier I: Manual Data, Manual Analysis; Tier II: Automated Data, Manual Analysis; and Tier III: Management by Exception. Most producers begin with Tier I systems and gradually add functionality over time. However, most of the value of production optimization occurs at the Tier II level. Therefore, producers should consider a Tier II or Tier III system.

Appendix

Figure Credits

All images are copyrighted and may not be reprinted, reproduced, or used in any way without the express written permission of the owner.

Figure		Owner	Web site
Frontis	Beam-lift units at work	The University of Texas at Austin, PETEX	www.utexas.edu/ce/petex
1	General arrangement of a reciprocating rod pump	The University of Texas at Austin, PETEX	www.utexas.edu/ce/petex
2	Rod pumping unit	Copyright © Weatherford International. All rights reserved.	www.weatherford.com
3	Early standard rig front arranged for pumping	The University of Texas at Austin, PETEX	www.utexas.edu/ce/petex
4	Early standard pumping rig driven by an electric motor drive	The University of Texas at Austin, PETEX	www.utexas.edu/ce/petex
5	Subsurface rod pump	The University of Texas at Austin, PETEX	www.utexas.edu/ce/petex
6	Steps of the pumping cycle for a conventional tubing pump	The University of Texas at Austin, PETEX	www.utexas.edu/ce/petex
7	Tubing pump and insert pump	Copyright © Weatherford International. All rights reserved.	www.weatherford.com
8	API insert pumps	The University of Texas at Austin, PETEX	www.utexas.edu/ce/petex
9	Plungers	Copyright © Weatherford International. All rights reserved.	www.weatherford.com
10	Gas separator process	The University of Texas at Austin, PETEX	www.utexas.edu/ce/petex
11	Sucker rod pump nomenclature	Copyright © Weatherford International. All rights reserved.	www.weatherford.com
12	Modern conventional pumping unit	Copyright © Weatherford International. All rights reserved.	www.weatherford.com

Figure		Owner	Web site
13	Parts of a conventional pumping unit	The University of Texas at Austin, PETEX	www.utexas.edu/ce/petex
14	Front-mounted geometry crank counterbalance unit	Copyright © Lufkin Industries Incorporated. All rights reserved.	www.lufkin.com
15	Phased crank counterbalance unit	Copyright © Weatherford International. All rights reserved.	www.weatherford.com
16	Beam balanced unit	Copyright © Weatherford International. All rights reserved.	www.weatherford.com
17	Long-stroke pumping unit	Copyright © Weatherford International. All rights reserved.	www.weatherford.com
18	Low-profile pumping unit	Copyright © Weatherford International. All rights reserved.	www.weatherford.com
19	Hydraulic pumping unit	Copyright © Weatherford International. All rights reserved.	www.weatherford.com
20	Sucker rods	Copyright © Weatherford International. All rights reserved.	www.weatherford.com
21	Sucker rod with an API pin	Copyright © Tenaris. All rights reserved.	www.tenaris.com
22	API sucker rods	Copyright © Tenaris. All rights reserved.	www.tenaris.com
23	Hollow sucker rod	Copyright © Tenaris. All rights reserved.	www.tenaris.com
24	Continuous rod installation	Copyright © Weatherford International. All rights reserved.	www.weatherford.com
25	Sucker rod guides	The University of Texas at Austin, PETEX	www.utexas.edu/ce/petex
26	Stacking rods on a rack off the ground prevents bending.	Copyright © Weatherford International. All rights reserved.	www.weatherford.com
27	Electric submersible pump	Copyright © Weatherford International. All rights reserved.	www.weatherford.com
28	Graph of efficiency versus speed for electric submersible pumps	Copyright © Weatherford International. All rights reserved.	www.weatherford.com
29	Schematic of centrifugal pump operation	The University of Texas at Austin, PETEX	www.utexas.edu/ce/petex
30	Impellers	Copyright © Weatherford International. All rights reserved.	www.weatherford.com
31	Radial flow and mixed flow stages	Copyright © Weatherford International. All rights reserved.	www.weatherford.com
32	ESP motor	Copyright © Weatherford International. All rights reserved.	www.weatherford.com

Figure		Owner	Web site
33	ESP motor seal section	Copyright © Weatherford International. All rights reserved.	www.weatherford.com
34	ESP gas separator	Copyright © Weatherford International. All rights reserved.	www.weatherford.com
35	ESP power cable	Copyright © Borets. All rights reserved.	www.borets.com
36	ESP wellhead	Copyright © Weatherford International. All rights reserved.	www.weatherford.com
37	Progressing cavity pump	Copyright © Weatherford International. All rights reserved.	www.weatherford.com
38	Cutaway view of a progressing cavity pump	Copyright © Weatherford International. All rights reserved.	www.weatherford.com
39	PC pump surface drivehead	Copyright © Weatherford International. All rights reserved.	www.weatherford.com
40	Cutaway view of an electric submersible progressing cavity pump	Copyright © Baker Hughes Incorporated. All rights reserved.	www.bakerhughes.com
41	Schematic of a PCP rotor, stator, and cavity	Copyright © Weatherford International. All rights reserved.	www.weatherford.com
42	PCP rotor in stator	Copyright © Weatherford International. All rights reserved.	www.weatherford.com
43	PCP rotor and stator pitches	Copyright © Weatherford International. All rights reserved.	www.weatherford.com
44	The rotor in a PC pump creates overlapping cavities.	Copyright © Weatherford International. All rights reserved.	www.weatherford.com
45	1:2 and 2:3 stator configurations	Copyright © Weatherford International. All rights reserved.	www.weatherford.com
46	Insert PC pump	Copyright © Weatherford International. All rights reserved.	www.weatherford.com
47	Electric drivehead for a PC pump	Copyright © Weatherford International. All rights reserved.	www.weatherford.com
48	Gas engine-driven drivehead for a PC pump	Copyright © Weatherford International. All rights reserved.	www.weatherford.com
49	Torque anchor	Copyright © Advantage Products Incorporated. All rights reserved.	www.advantageproductsinc.com
50	Pump efficiency versus pump lift	Copyright © Weatherford International. All rights reserved.	www.weatherford.com
51	Composite flow tee with rod-gripping feature	Copyright © Weatherford International. All rights reserved.	www.weatherford.com
52	Gas lift	Copyright © Weatherford International. All rights reserved.	www.weatherford.com

Figure		Owner	Web site
53	Natural flow and gas lift	The University of Texas at Austin, PETEX	www.utexas.edu/ce/petex
54	Unloading sequence	The University of Texas at Austin, PETEX	www.utexas.edu/ce/petex
55	Bellows gas-lift valve	Copyright © Weatherford International. All rights reserved.	www.weatherford.com
56	Gas-lift mandrel	Copyright © Weatherford International. All rights reserved.	www.weatherford.com
57	Gas-lift running tool	Copyright © Weatherford International. All rights reserved.	www.weatherford.com
58	Single perforation for gas lift	The University of Texas at Austin, PETEX	www.utexas.edu/ce/petex
59	Closed configuration, intermittent gas lift	Copyright © Weatherford International. All rights reserved.	www.weatherford.com
60	System deliverability curve	Copyright © Weatherford International. All rights reserved.	www.weatherford.com
61	Plunger lift	Copyright © Weatherford International. All rights reserved.	www.weatherford.com
62	Plunger lift surface assembly	Copyright © Weatherford International. All rights reserved.	www.weatherford.com
63	Progressive (staged) plunger lift	Copyright © Weatherford International. All rights reserved.	www.weatherford.com
64	Conventional plungers	Copyright © Weatherford International. All rights reserved.	www.weatherford.com
65	Continuous flow plungers	Copyright © Weatherford International. All rights reserved.	www.weatherford.com
66	Subsurface assembly	Copyright © Weatherford International. All rights reserved.	www.weatherford.com
67	Plunger-lift lubricator	Copyright © Weatherford International. All rights reserved.	www.weatherford.com
68	Plunger-lift controller	Copyright © Weatherford International. All rights reserved.	www.weatherford.com
69	Critical gas rate and related equations	The University of Texas at Austin, PETEX	www.utexas.edu/ce/petex
70	Internal capillary-injection string	Copyright © Weatherford International. All rights reserved.	www.weatherford.com
71	Hydraulic-lift system	Copyright © Weatherford International. All rights reserved.	www.weatherford.com
72	Surface equipment for hydraulic pumping	The University of Texas at Austin, PETEX	www.utexas.edu/ce/petex

Figure		Owner	Web site
73	Hydraulic jet pump	Copyright © Weatherford International. All rights reserved.	www.weatherford.com
74	Hydraulic piston pump operation	Copyright © Weatherford International. All rights reserved.	www.weatherford.com
75	Free pump casing return	Copyright © Weatherford International. All rights reserved.	www.weatherford.com
76	Free pump parallel return	Copyright © Weatherford International. All rights reserved.	www.weatherford.com
77	Tubing-conveyed pumps	Copyright © Weatherford International. All rights reserved.	www.weatherford.com
78	Coiled tubing jet pump	Copyright © Weatherford International. All rights reserved.	www.weatherford.com
79A	Sliding sleeve wireline application	Copyright © Weatherford International. All rights reserved.	www.weatherford.com
79B	Gas-lift mandrel application	Copyright © Weatherford International. All rights reserved.	www.weatherford.com
80	Schematic of a hydraulic jet pump	Copyright © Weatherford International. All rights reserved.	www.weatherford.com
81	Engine-type section of a hydraulic piston pump	Copyright © Weatherford International. All rights reserved.	www.weatherford.com
82	Self-contained surface unit of a hydraulic piston pump	The University of Texas at Austin, PETEX	www.utexas.edu/ce/petex
83	Fluid supply plant of a hydraulic piston pump	The University of Texas at Austin, PETEX	www.utexas.edu/ce/petex
84	Main categories of production optimization	Copyright © Weatherford International. All rights reserved.	www.weatherford.com
85	Subsurface flowmeter	Copyright © Weatherford International. All rights reserved.	www.weatherford.com
86	Watercut meter	Copyright © Weatherford International. All rights reserved.	www.weatherford.com
87	Fiber-optic cable	The University of Texas at Austin, PETEX	www.utexas.edu/ce/petex
88	Schematic of a SCADA system	Copyright © Weatherford International. All rights reserved.	www.weatherford.com
89	Example of the user interface for a desktop intelligence software system	Copyright © Weatherford International. All rights reserved.	www.weatherford.com
90	Example schematic of an integrated system	Copyright © Weatherford International. All rights reserved.	www.weatherford.com

Glossary

A

aeration *n*: the injection of air or gas into a liquid. In the oil industry, a common form of aeration is the injection of natural gas into reservoir liquids standing in a well. Aeration with natural gas reduces the density of the liquids and allows declining reservoir pressure to lift the liquids. See *gas lift*.

American Petroleum Institute (API) *n*: an oil trade organization (founded in 1920) that is the leading standardizing organization for oilfield drilling and producing equipment. Headquartered in Washington D.C., it publishes materials concerning exploration and production, petroleum measurement, marine transportation, marketing, pipelining, refining, safety and fire protection, storage tanks, valves, training, health and environment, policy, and economic studies. Address: 1220 L Street NW; Washington, DC 20005; 202-682-8000; www.api.org.

API *abbr*: American Petroleum Institute.

B

ball-and-seat check valve *n*: a device used to restrict fluid flow to one direction. It consists of a polished sphere, or ball, usually of metal, and an annular piece, the seat, ground and polished to form a seal with the surface of the ball. Gravitational force or the force of a spring holds the ball against the seat. Flow in the direction of the force is prevented, while flow in the opposite direction overcomes the force and unseats the ball.

ball valve *n*: a flow-control device employing a ball with a rotating mechanism to open or close the tubing.

BHP *abbr*: bottomhole pressure.

BHT *abbr*: bottomhole temperature.

blowout preventer (BOP) *n*: one of several valves installed at the wellhead to stop (prevent) the escape of pressure either in the annular space between the casing and the drill pipe or in open hole (i.e, hole with no drill pipe) during drilling or completion operations. Blowout preventers on land rigs are normally located beneath the rig at, or slightly below, the land's surface; on jackup or platform rigs, at the water's surface; and on floating offshore rigs, on the seafloor.

BOP *abbr*: blowout preventer.

bottomhole pressure (BHP) *n*: 1. the pressure at the bottom of a borehole. It is caused by the hydrostatic pressure of the wellbore fluid and, sometimes, by any back-pressure held at the surface, as when the well is shut in with blowout preventers. When mud is being circulated, bottomhole pressure is the hydrostatic pressure plus the remaining circulating pressure required to move the mud up the annulus. 2. the pressure in a well at a point opposite the producing formation, as recorded by a pressure-tight container mechanism referred to as a bottomhole pressure bomb.

bottomhole temperature (BHT) *n*: temperature measured in a well at a depth at the midpoint of the thickness of the producing zone.

brine *n*: water that has a large quantity of salt, especially sodium chloride, dissolved in it; salt water.

C

centipoise (cp) *n*: one-hundredth of a poise; a measure of a fluid's viscosity, or resistance to flow.

centrifugal pump *n*: a dynamic pump with an impeller or rotor, an impeller shaft, and a casing, which discharges fluid and boosts pressure by centrifugal force. Because the pump is dynamic rather than positive displacement, pump performance is dependent upon maintaining high rotational velocity. An electric submersible pump is a multistage centrifugal pump.

check valve *n*: a valve that permits fluid to flow in one direction only. If the gas or liquid starts to reverse, the valve automatically closes, preventing reverse movement. Sometimes referred to as a one-way valve.

coiled tubing *n*: a continuous string of flexible steel tubing, often hundreds or thousands of feet long, that is wound onto a reel, often dozens of feet in diameter. The reel is an integral part of the coiled-tubing unit, which consists of several devices that ensure the tubing can be safely and efficiently inserted into the well from the surface. Because tubing can be lowered into a well without having to make up joints of tubing, running coiled tubing into the well is faster and less expensive than running conventional tubing. Rapid advances in the use of coiled tubing make it a popular way in which to run tubing into and out of a well. Also called reeled tubing.

continuous gas lift *n*: see *gas lift*.

continuous rod *n*: a single length of rod that is the mechanical link between the surface pumping unit or surface drivehead and a subsurface pump. Continuous rod is spooled into and out of the well. A sucker rod pin is welded to the bottom of the continuous rod for connection to the subsurface pump. A sucker rod pin or polished rod pin is welded to the top of the continuous rod for connection to polished rod to provide a seal surface for pressure containment. Continuous rod can have a round cross section or a non-round cross section to facilitate spooling. Rods with non-round cross-sections are not used in rotating rod (PCP) applications.

continuous sucker rod *n*: see *continuous rod*.

counterbalance weight *n*: a weight applied to compensate for existing weight or force. On pumping units in oil production, counterweights are used to offset the weight of the rod string and fluid on the upstroke of the pump and the weight of the rods on the downstroke.

counterweight *n*: see *counterbalance weight*.

cp *abbr*: centipoise.

critical velocity *n*: (as used in the oil and gas industry) the gas velocity at which a drop of liquid is suspended within a vertical flow of gas, i.e., the downward gravitational force on a drop of liquid is balanced by the fluid drag from the upward gas flow. Critical velocity is used to determine whether or not gas flow will remove liquids from a producing well.

D

dead string *n*: a velocity string with no internal flow. All flow occurs through the annulus around the velocity string. See *velocity string*.

decline curve *n*: the graphical representation of production rate versus time, typically for an individual well.

deviation *n*: departure of the wellbore from the vertical, measured by the horizontal distance from the rotary table to the target. The amount of deviation is a function of the drift angle and hole depth. The term is sometimes used to indicate the angle from which a bit has deviated from the vertical during drilling.

diffuser *n*: a device that uses part of the kinetic energy of a fluid passing through a machine by gradually increasing the cross section of the channel or chamber through which it flows so as to decrease its speed and increase its pressure.

diluent *n*: liquid added to dilute or thin a solution.

E

elastomer *n*: a synthetic rubber made from polymers that has the elastic properties of natural rubber; packers (sealing elements) in blowout preventers and downhole packers are often made of elastomer. The term is formed by combining part of the word elastic and part of the word polymer—i.e., elast(ic) and poly(mer).

electric submersible progressing cavity pump (ESPCP) *n*: a progressing cavity pump (PCP) driven by a subsurface ESP motor.

electric submersible pump (ESP) *n*: an electrically-powered, multistage centrifugal pump. In each stage, an impeller accelerates fluid into a diffuser which decelerates the fluid in a controlled manner to convert fluid energy into increased pressure. Electrical power from the surface is conducted to a subsurface electric motor in the pump assembly by a cable attached to the tubing. The pump assembly also typically includes a seal and bearing assembly and can include gas separation and other fluid conditioning features.

electric submersible pumping *n*: a form of artificial lift that utilizes an ESP lowered to a depth in the well that submerges it into the liquids produced by the well.

enhanced oil recovery (EOR) *n*: 1. the introduction of artificial drive and displacement mechanisms into a reservoir to produce a portion of the oil unrecoverable by primary recovery methods. To restore formation pressure and fluid flow to a substantial portion of a reservoir, fluid or heat is introduced through injection wells located in rock that has fluid communication with production wells. 2. the use of certain recovery methods that not only restore formation pressure but also improve oil displacement or fluid flow in the reservoir. These methods include chemical flooding, gas injection, and thermal recovery.

EOR *abbr*: enhanced oil recovery.

ESP *abbr*: electric submersible pump.

ESPCP *abbr*: electric submersible progressing cavity pump.

F

flapper valve *n*: a type of check valve in a pipe or a line that has a hinged closure mechanism (a flapper) and that allows fluid flow in one direction but shuts it off in the other direction.

fluid *n*: a substance that flows and yields to any force tending to change its shape. Liquids and gases are considered fluids. Fluids assume the shape of the container in which they are placed.

free gas *n*: a hydrocarbon that exists in the gaseous phase at reservoir pressure and temperature and remains a gas when produced under normal conditions.

G

gas-cap drive *n*: drive energy supplied naturally (as a reservoir is produced) by the expansion of the gas cap. In such a drive, the gas cap expands to force oil into the well and to the surface.

gas-cap reservoir *n*: see *gas-cap drive*.

gas drive *n*: 1. the use of the energy that arises from the expansion of compressed gas in a reservoir to move crude oil to a wellbore. The gas can be in an overriding cap or in solution. Also called depletion drive. See *gas-cap drive*. 2. gas drives are also used as a form of secondary recovery in which gas is injected into wells to sweep remaining oil to a producing well.

gas interference *n*: a condition in which gas partially displaces liquid in a pump resulting in less fluid movement. Some of the pump energy and motion is expended compressing and expanding gas rather than displacing fluids. See *gas lock*.

gas lift *n*: the process of raising or lifting fluid from a well by injecting gas. In conventional gas lift, injected gas aerates the produced fluid in order to lower the weight of the produced fluid column to the point that well pressure can displace the fluid column to the surface. Multiple gas lift valves are used to gradually lower the fluid column weight in a stepwise sequence to minimize the required injection gas pressure. In intermittent gas lift, gas in the annulus is suddenly released into the production tubing at high velocity to displace liquids above the injection point to the surface.

gas-lift valve (GLV) *n*: a device for injection of gas into production fluids. Gas lift valves are installed on or in gas-lift mandrels to provide flow communication between the casing-tubing annulus and the tubing string of a gas-lift well. Gas lift valves contain check arrangements to allow flow in one direction only. Tubing and casing pressures cause the valve to open and close at preset pressures.

gas-liquid ratio (GLR) *n*: a measure of the volume of gas produced with liquid.

gas lock *n*: 1. a condition in which the working chamber(s) of a pump are filled with gas such that the pumping motion compresses and expands gas without moving fluids through the pump. In a rod pump, dissolved gas can be released from solution during the upstroke of the plunger and appears as free gas between the valves. During subsequent upstrokes, if the gas pressure remains higher than the pressure of the production fluids below the standing valve, the standing valve remains shut, and no fluid enters the pump. During the downstroke, if the gas pressure remains less than the pressure in the production tubing, the traveling valve remains shut and no fluid exits the pump. 2. a device fitted to the gauging hatch on a pressure tank that enables manual dipping and sampling without loss of vapor.

3. a condition that can occur when gas-cut mud is circulated by the mud pump. The gas breaks out of the mud, expands, and works against the operation of the piston and valves.

GLR *abbr*: gas-liquid-ratio.

GLV *abbr*: gas-lift valve.

H

heavy oil *n*: oil composed mainly of heavy ends., with API gravity between 10° and 22.3° (Reference: World Petroleum Conference)

hertz (Hz) *n*: a unit in the metric system used to measure frequency in cycles per second.

hold-down *n*: a mechanical arrangement that prevents the upward movement of certain pieces of equipment installed in a well. A sucker rod pump can use a mechanical hold-down for attachment to a seating nipple.

hp *abbr*: horsepower.

HP *abbr*: hydrostatic pressure.

hydraulic jet pump *n*: a specialized form of hydraulic pump used in artificial lift. The main working parts of a hydraulic jet pump include a nozzle, throat, and diffuser. The nozzle converts the high-pressure, low-velocity energy of the power fluid to high-velocity, low-pressure energy. The power fluid is then mixed with the low-pressure pump intake fluid in the throat to produce a low-pressure stream with a velocity less than that of the nozzle exit but a high velocity nevertheless. The velocity energy of this mixed stream is then converted to static pressure in the diffuser to provide the pressure necessary to lift fluid from the well. The power fluid is typically oil or water.

hydraulic lift *n*: see *hydraulic pumping*.

hydraulic piston pump *n*: a subsurface hydraulic pump used in artificial lift consisting of an engine section that drives a reciprocating pump to produce fluids. Pressurized power fluid from the surface causes a piston in the engine section to stroke. The engine piston is coupled to a piston in the pump section which strokes to displace produced fluids. The engine section reverses direction at the end of the stroke causing the pump section to likewise reciprocate. In this way, the hydraulic piston pump has similar performance characteristics to a rod pump except without a rod string.

hydraulic pump *n*: 1. a subsurface pump that gets its energy from pressurized liquid from the surface to lift produced liquids from wells. See *hydraulic pumping*. 2. a device that creates pressure on a fluid, usually special hydraulic fluid, to move the fluid.

hydraulic pumping *n*: a method of pumping oil liquids from wells by using a downhole subsurface hydraulic pump. Hydraulic pumps can be deployed and retrieved free (gravity or pumped into place), using wireline, or with conventional-tubing-conveyed. Hydraulic systems can be open loop in which expended power fluid is comingled with produced fluids or returned to the surface in a closed loop, segregated flow path.

hydrostatic pressure (HP) *n*: the force exerted by a body of fluid at rest. It increases directly with the density and the depth of the fluid and is expressed in many different units, including pounds per square inch or kilopascals.

The hydrostatic pressure of fresh water is 0.433 pounds per square inch per foot (9.792 kilopascals per metre) of depth. In drilling, the term refers to the pressure exerted by the drilling fluid in the wellbore. In a water drive field, the term refers to the pressure that furnishes the primary energy for production.

Hz *abbr*: hertz, (i.e., cycles per second).

I

impeller *n*: a bladed rotor; its rotation imparts centrifugal motion to a fluid (e.g., the rotor of a centrifugal pump).

insert pump *n*: a pump that is run into the well as a complete unit. Examples include insert rod pumps and insert PC pumps.

intermittent gas lift *n*: see *gas lift*.

K

k *abbr*: permeability.

L

line loss *n*: 1. the reduction in the quantity of natural gas flowing through a pipeline that results from leaks, venting, and other physical and operational circumstances. 2. the reduction in (or the loss of) electrical energy (voltage) that occurs when electricity flows through a conductor (a line).

liquid loading *n*: the accumulation of liquids within a wellbore that inhibits gas inflow.

M

mandrel *n*: 1. a cylindrical bar, spindle, or shaft around which other parts are arranged or attached or that fits inside a cylinder or tube. 2. a tube or hollow member in the production tubing containing a gas lift valve for injection of gas into the production fluids for the purpose of gas lift. See *gas lift* and *side pocket mandrel*. 3. a tube or hollow member in the production tubing containing a gauge for subsurface measurements.

multistage centrifugal pump *n*: a centrifugal pump that develops pressure by means of impeller-diffuser pairs (pump stages) operating in series. See *electric submersible pumping*.

N

natural flow *n*: fluid production from a subsurface reservoir to the surface without assistance from artificial means. See *gas lift* and *aeration*.

O

overtravel *n*: the difference between the subsurface rod pump piston stroke and the surface polished rod stroke, usually the result of system harmonic response. Also, over-travel.

oxidation *n*: a chemical process in which oxygen combines with a compound and causes the compound to lose electrons and gain a more positive charge. For example, when exposed to air (which contains about 21% oxygen), iron rusts, which means that part of the iron chemically combines with the oxygen in the air and oxidizes, or becomes rusty. The rust is typically a red iron oxide, which gives iron rust its characteristic color.

P

packer *n*: a piece of downhole equipment that consists of a sealing device, a holding or setting device, and an inside passage for fluids. It is used to block the flow of fluids through the annular space between pipe and the wall of the wellbore by sealing off the space between them. In production, it is usually made up in the tubing string some distance above the producing zone. A packing element expands to prevent fluid flow except through the packer and tubing. Packers are classified according to configuration, use, and method of setting and whether or not they are retrievable (that is, whether they can be removed when necessary, or whether they must be milled or drilled out and thus destroyed).

paraffin *n*: a saturated aliphatic hydrocarbon having the formula C_nH_{2n+2} (e.g., methane, CH_4; ethane, C_2H_6). Heavier paraffin hydrocarbons (i.e., $C_{18}H_{38}$) form a waxlike substance that is called paraffin. These heavier paraffins often accumulate on the walls of tubing and other production equipment, restricting or stopping the flow of the desirable lighter paraffins.

PCP *abbr*: progressing cavity pump. Also, PC pump.

PC pump *abbr*: progressing cavity pump. Also, PCP.

perforate *v*: to pierce the casing wall and cement of a wellbore to provide holes through which formation fluids can enter or to provide holes in the casing so that materials can be introduced into the annulus between the casing and the wall of the borehole. Perforating is accomplished by lowering into the well a perforating gun, or perforator, which fires electrically detonated bullets or shaped charges.

permeability (k) *n*: 1. a measure of the ease with which a fluid flows through the connecting pore spaces of rock or cement. The unit of measurement is the millidarcy. 2. fluid conductivity of a porous medium. 3. ability of a fluid to flow within the interconnected pore network of a porous medium. 4. on offshore drilling rigs, the percentage of a given space in a vessel that can be occupied by water. 5. the property of a magnetizable substance that determines the degree to which it modifies the magnetic lines of force (the flux) in the region occupied by the substance in a magnetic field; specifically, it is the ratio of the induction to the magnetizing force in the substance.

permeable *adj*: allowing the passage of fluid. See *permeability*.

PI *abbr*: productivity index.

piston *n*: a cylindrical sliding piece that is moved by or that moves against fluid pressure within a confining cylindrical vessel.

plunger *n*: 1. a free piston that is motivated by produced and/or injected gas to displace liquid above the plunger to the surface. See *plunger lift*. 2. a basic component of the rod pump that serves to draw well fluids into the pump. See *rod pump*. 3. the rod that serves as a piston in a reciprocating pump. 4. the device in a fuel-injection unit that regulates the amount of fuel pumped on each stroke.

plunger lift *n*: a method of artificial lift in which a plunger travels up and down inside production tubing to move slugs of liquid to the surface. Produced gas and/or injected gas below the plunger provides the lifting energy. A conventional plunger cycles in response to a timer or surface controller which temporarily shuts in the well to allow the plunger to fall and for pressure to build below the plunger. Continuous flow plungers have internal bypass features which allow continuous operation without shutting in the well.

polished rod *n*: a rod at the topmost portion of a rod string with a smooth surface for contact with a dynamic seal in the surface stuffing box in order to isolate well pressure and prevent leakage of well fluids. Polished rods are used in reciprocating rod lift and rotating rod applications (PCPs).

positive-displacement pump *n*: a reciprocating or a rotary pump that moves a measured quantity of liquid with each stroke or rotation of the pump.

power fluid *n*: in subsurface hydraulic pumping, the liquid that is pressurized and pumped into the well to power the subsurface pump. The power fluid is typically processed well fluid or water.

productivity index (PI) *n*: a well-test measurement indicative of the amount of oil or gas a well is capable of producing. It may be expressed as—

$$PI = q \div (P_s - P_f)$$

where—

PI = productivity index (barrels/day or thousand cubic feet/day per pounds per square inch of pressure differential)

q = rate of production (barrels/day or thousand cubic feet/day)

P_s = static bottomhole pressure (pounds per square inch)

P_f = flowing bottomhole pressure (pounds per square inch).

progressing cavity pump (PC pump, PCP) *n*: a positive displacement pump consisting of a corkscrew-shaped rotor turning inside a helical cavity stator such that sealed cavities of fluid progress from the pump inlet to the outlet. Conventional PC pumps are rotated by rod strings from a surface drivehead. PC pumps can also be driven by subsurface electric or hydraulic motors.

R

reciprocating motion *n*: back-and-forth or up-and-down movement, such as that of a piston in a cylinder.

reciprocating pump *n*: see *rod pump*.

reciprocating rod lift *n*: a method of artificial lift in which a surface pumping unit reciprocates a rod string connected to a subsurface pump located at or near the bottom of the well to lift the well fluids to the surface.

reciprocation *n*: a back-and-forth or up-and-down movement (as the movement of a piston in an engine or pump).

revolutions per minute (rpm) *n*: the number of full rotations completed in one minute around a fixed axis. Generally used in measuring the rotational speed of a mechanical device.

rod pump *n*: a pump consisting of a piston that reciprocates back and forth or up and down in a cylinder (barrel). The cylinder is equipped with an inlet (suction) standing valve and outlet (discharge) traveling valves. On the intake stroke, the standing valve opens and fluid is drawn into the cylinder. On the discharge stroke, the standing valve closes, the traveling valve opens, and fluid is forced out of the cylinder.

rod string *n*: the entire length of sucker rods or continuous rod that serves as a mechanical link from the beam pumping unit on the surface to the rod pump near the bottom of the well.

rpm *abbr*: revolutions per minute.

S

SCADA *abbr*: Supervisory Control and Data Acquisition systems.

screw pump *n*: see *progressing cavity pump*.

side pocket mandrel (SPM) *n*: a hollow member in the production tubing with a side cavity that contains a gas lift valve for injection of gas for gas lift. Gas lift valves can be run, landed in the SPM, and retrieved using wireline or coiled tubing without pulling tubing or the SPM. See *mandrel*, *gas lift*, and *gas-lift valve*.

slick line *n*: see *wireline*.

slippage *n*: internal fluid leakage within a pump, equal to volumetric inefficiency. Some slippage is necessary in most mechanical pumps in order to provide lubrication between sliding surfaces and to provide cooling.

solution gas drive *n*: a source of natural reservoir energy in which the solution gas coming out of the oil expands to force the oil into the wellbore.

SPM *abbr*: side pocket mandrel.

SSSV *abbr*: subsurface safety valve.

standing valve *n*: a fixed ball-and-seat valve at the lower end of a rod pump. During the upstroke, the standing valve allows inflow of well fluids into the rod pump. During the downstroke, the standing valve supports the produced fluid column and prevents the well fluids from escaping back into the reservoir. Compare *traveling valve*.

static pressure *n*: 1. the stationary or line pressure existing in a vessel or pipe. 2. the pressure exerted by a fluid on a surface that is at rest in relation to the fluid. 3. the pressure exhibited at the surface or point downhole during the time the well is shut in. 4. surface or bottomhole pressure after sufficient time has elapsed for the pressure to become stable.

subsurface safety valve (SSSV) *n*: a device installed in the tubing of the producing well to shut in the flow of production if the flow exceeds a preset rate. Tubing safety valves are widely used in offshore wells to prevent pollution if the wellhead fails for any reason. Also called a tubing safety valve.

sucker rod *n*: a special steel pumping rod. Several rods screwed together make up the mechanical link from the beam pumping unit on the surface to the rod pump at the bottom of a well. Sucker rods are threaded on each end and manufactured to dimension standards and metal specifications set by the petroleum industry. Standard lengths are 25 feet or 30 feet (7.6 metres or 9.1 metres); common sizes are from ⅝ inch to 1½ inch (12 millimetres to 38 millimetres).

sucker rod pump *n*: see *rod pump*.

sucker rod pumping *n*: see *reciprocating rod lift*.

Supervisory Control and Data Acquisition (SCADA) *n*: a system to collect and communicate sensor data from the wellsite to remote locations. SCADA systems can be configured to transmit data from remote locations to controllers at the wellsite.

surfactant *n*: a soluble compound that reduces the surface tension of substances. The use of surfactants permits the thorough surface contact or mixing of substances that ordinarily remain separate. In artificial lift applications, surfactants are used to turn liquids to foam in the presence of fluid turbulence so the liquids can be produced to the surface using gas flow velocity. Surfactants are used to improve the mixing of compounds in drilling mud and in water and chemical flooding.

synchronous speed *n*: in a two-pole induction motor, the maximum speed at which such a motor is capable of operating. In a 60-Hz system, the synchronous speed is 3,600 revolutions per minute (rpm). Synchronous speed can be stated as the equation—

$$\text{synchronous speed (rpm)} = \frac{120 \times \text{frequency (hertz)}}{\text{number of poles}}$$

T

tensile strength *n*: 1. the actual stress level at which a material catastrophically fails in tension. 2. the maximum allowable tensile stress rating for a material to avoid catastrophic failure in tension.

tensile stress *n*: stress (force per unit area) developed within a material in response to tensile forces aligned in opposite directions as if to pull the material apart.

torque *n*: the turning force that is applied to a shaft or other rotary mechanism to cause it to rotate or tend to do so. Torque is expressed as a force acting at a distance from a centerline (foot-pounds, newton-metres).

traveling valve *n*: a ball-and-seat valve within the plunger of a rod pump. During the plunger upstroke, the traveling valve supports and lifts the fluid column. During the downstroke, the traveling valve opens to allow the plunger to descend through the fluid in the pump so fresh fluid flows above the traveling valve. Compare *standing valve*.

tubing pump *n*: a rod pump in which the barrel is made up as part of the tubing (e.g. tubing rod pumps and tubing conveyed PC pumps).

U

undertravel *n*: the difference between the surface polished rod stroke and the subsurface rod pump piston stroke, usually the result of system friction or system harmonic response. Also, under-travel.

unloading the well *vp*: removing fluid from the tubing in a well to lower the bottomhole pressure to induce the well to flow. Common unloading technologies include swabbing, gas lift, and jet pumping.

V

velocity string *n*: a length of tubing inserted into production tubing or casing to provide reduced flow area for the purpose of increasing the velocity of produced gas. Flow can occur through the velocity string or through the annulus around the velocity string. See *dead string*.

venturi effect *n*: the drop in pressure resulting from the increased velocity of a fluid as it flows (Bernoulli effect).

venturi nozzle *n*: a nozzle that provides high velocity flow to create a venturi effect pressure drop in order to draw in production fluids from an adjacent intake.

viscosity *n*: a measure of the resistance of a fluid to flow. Resistance is brought about by the internal friction resulting from the combined effects of cohesion and adhesion. The viscosity of petroleum products is commonly expressed in terms of the time required for a specific volume of the liquid to flow through an orifice of a specific size at a given temperature.

viscous *adj*: having a high resistance to flow.

W

water drive *n*: the pressurization and displacement of hydrocarbons in a reservoir from an aquifer adjacent to the reservoir (edgewater drive) or below the reservoir (bottomwater drive). Bottomwater drive is considered more efficient because of more contact area with the reservoir.

wireline *n*: 1. a high strength wire for running in and retrieving tools from the well. Also called, *slick line*. 2. braded electrical cable for running, retrieving, and powering subsurface tools.

Review Questions

WELL SERVICING AND WORKOVER

Lesson 5: Artificial Lift Methods

Multiple Choice

Pick the *best* answer from the choices and place the letter of that answer in the blank provided.

______ 1. A ______ PC pump utilizes a rod string to transfer torque to the downhole pump.

A. balanced
B. closed installation
C. conventional
D. pressure-actuated

______ 2. An electric submersible pump assembly must be connected to the surface equipment by a—

A. jointed rod.
B. insulated tubing string.
C. power cable.
D. fiber-optic cable.

______ 3. Continuous gas lift is most applicable to wells in reservoirs with substantial amounts of remaining energy and—

A. no abrasive solids.
B. low bottomhole pressure.
C. low to moderate amounts of produced water.
D. high productivity indexes.

______ 4. Pump flow rate is usually kept below the maximum possible flow rate for that well. This is done primarily—

A. to prevent formation damage.
B. to reduce corrosion and the destructive effects of CO_2.
C. to conform to noise restrictions.
D. because high-volume lift technologies are so complicated.

______ 5. When gas lift is initiated, all liquid in the annulus above the lowest gas-lift valve is removed, a process known as—

A. unloading.
B. pulsing.
C. underbalancing.
D. snubbing.

_____ 6. The most commonly used method of artificial lift is—

A. electric submersible pumping.
B. plunger lift.
C. reciprocating rod lift.
D. hydraulic lift.

_____ 7. When selecting an artificial lift system, it is important to—

A. review well construction and fluid composition beforehand.
B. determine target pressure and production capacity
C. eliminate lift technologies that are not viable.
D. all of the above.

_____ 8. The purpose of a SCADA system is to—

A. collect sensor data and communicate it to a remote location.
B. assist in the initial design and planning of an artificial lift system.
C. transfer energy to subsurface pumping devices.
D. automatically shut in a well if there is a loss of power.

_____ 9. In the majority of wells, over time—

A. the natural flow of hydrocarbons slowly increases and then abruptly stops.
B. water in the produced fluids is steadily replaced by oil.
C. the initial rate of fluid production does not change significantly.
D. formation pressure drops to the point that artificial lift is necessary.

_____ 10. One way to reduce up-front capital expenses when setting up an artificial-lift system is to—

A. use non-metallic components where possible.
B. lease lift equipment instead of buying it.
C. make extensive use of automated data collection and analysis tools.
D. use gas combustion engines to power lift systems.

_____ 11. Reciprocating rod lift systems account for approximately _______ of all artificial lift systems in operation.

A. 80%
B. 40%
C. 60%
D. 20%

_____ 12. Foam lift can be used effectively in conjunction with _______ to extend the life of the well.

A. plunger lift
B. velocity strings and dead strings
C. intermittent gas lift
D. electric submersible pumps

______ 13. Reciprocating rod systems should not be considered—

A. to produce heavy oil.
B. for reservoirs with low rates of production.
C. where water is present in the produced fluids.
D. for offshore applications.

______ 14. The presence of _______ will shorten the lifespan of ESP components.

A. gas in emulsion
B. produced water
C. particulate matter
D. heavy hydrocarbons

______ 15. In a reciprocating rod system, the _______ transfers the motion of the surface pumping unit to the subsurface pump.

A. venturi nozzle
B. check valve
C. impeller
D. rod string

______ 16. Conventional beam pumping units are valued for their—

A. applicability for highly deviated wells.
B. ability to run with no external power source.
C. reliability and ruggedness.
D. high-volume production capability.

______ 17. _______ systems are the most versatile artificial lift technology.

A. Plunger lift
B. Progressing cavity pumping
C. Hydraulic lift
D. Gas lift

______ 18. _______ is often the most effective way to improve production.

A. Increasing pumping rates
B. Increasing injection rates
C. Changing lift systems
D. Field-wide production optimization

______ 19. _______ is one of the preferred technologies for high-volume offshore oil production.

A. Electric submersible pumping
B. Reciprocating rod lift
C. Foam lift
D. Progressing cavity pump lift

______ 20. In general, reciprocating rod applications are limited by—

A. complexity of maintenance.
B. inefficiency when pumping light oil.
C. issues related to the rod string.
D. inability to lift heavy oil.

______ 21. Most hydrocarbons are lifted to the surface using—

A. rod pumping.
B. gas lift.
C. electric submersible pumps.
D. plunger lift.

______ 22. Gas lift installations usually include—

A. subsurface centrifugal pumps.
B. surfactant delivered through capillary strings.
C. multiple injection points.
D. a venturi nozzle.

______ 23. An electric submersible pump consists of an assembly of—

A. synchronized pistons.
B. automatic analysis tools.
C. bellows-type gas-injection valves.
D. impellers and diffusers.

______ 24. Rod-driven PC pumping systems are known for—

A. pumping heavy fluids ineffectively.
B. having relatively high initial costs.
C. representing the most forgiving lift technology.
D. having high operating efficiency.

______ 25. In a progressing cavity pump, a rotor turns inside—

A. an impeller.
B. a diffuser.
C. a standing valve.
D. a stator.

______ 26. A(n) ______ system operates using only the energy of the formation.

A. intermittent gas lift
B. plunger lift
C. progressing cavity pump
D. hydraulic lift

_____ 27. Compared to air, natural gas is preferred in gas-lift systems because it provides more lift and—

A. does not cause oxidation.
B. prevents formation damage.
C. dissolves into the produced liquids.
D. removes scale buildup.

_____ 28. Production optimization is—

A. reducing friction in either the annulus or production tubing.
B. managing the production of hydrocarbons over the life of the well.
C. designing and applying lift technology in order to minimize the initial investment.
D. filtering and cleaning the fluids produced from a well or field.

_____ 29. The lift technology with the lowest cost per barrel lifted for low-volume applications is—

A. electric submersible pumping.
B. foam lift.
C. plunger lift.
D. hydraulic lift.

_____ 30. Progressing cavity pumps require special elastomer compounds when exposed to—

A. aromatic gases or CO_2.
B. fluid slippage.
C. notches and stress risers.
D. gas locking.

_____ 31. The presence of sand and particulate matter does not negatively affect—

A. reciprocating rod lift systems.
B. gas-lift systems.
C. electric submersible pumping systems.
D. plunger-lift systems with brush plungers.

_____ 32. A progressive plunger lift system includes—

A. two or more plungers acting in series within the same tubing.
B. multiple unloading valves.
C. a corkscrew-shaped rotor turning inside a cavity.
D. power fluid pumped down the production tubing, pushing mingled production and power fluids up the annulus.

_____ 33. In a foam lift system, chemicals are introduced to reduce—

A. turbulence.
B. surface tension of produced fluids.
C. sloughing into the hole.
D. acidity.

______ 34. In a vertical gas flow, the point at which the forces tending to lift a droplet of liquid are balanced with the gravitational force on that droplet is known as—

A. critical velocity.
B. the displacement point.
C. positive uplift.
D. gas interference.

______ 35. A hydraulic lift system includes extensive equipment at the surface to—

A. hold the weight of the subsurface pump.
B. dispense surfactants into the well.
C. deliver electricity to the downhole components.
D. separate and clean the power fluid.

______ 36. In a plunger lift system, a _______ is mounted at the bottom of the hole.

A. venturi nozzle
B. bumper spring assembly
C. capillary tube array
D. magnet motor

______ 37. In a reciprocating rod system, on the downstroke of the pump, the _______ allows fluid to pass through the hollow core of the plunger so the plunger can fall into position.

A. traveling valve
B. foot valve
C. stationary valve
D. counterbalance valve

______ 38. _______ is especially effective for producing viscous and abrasive fluids.

A. Reciprocating rod lift
B. Progressing cavity pumping
C. Foam lift
D. Electric submersible pumping

______ 39. A challege of hydraulic jet pump systems is their—

A. narrow range of applications.
B. vulnerability to overheating.
C. high fluid velocity.
D. high energy requirement.

______ 40. Compared to other systems, electric submersible pumping systems are limited by—

A. their inability to produce high volumes of fluids.
B. a narrow operating speed range or pump turndown ratio.
C. their inefficiency in horizontal wells.
D. harmonic vibration damage.

Index

Answer Key

WELL SERVICING AND WORKOVER

Lesson 5: Artificial Lift Methods

1. C
2. C
3. D
4. A
5. A
6. C
7. D
8. A
9. D
10. B
11. A
12. B
13. D
14. C
15. D
16. C
17. C
18. D
19. A
20. C
21. C
22. C
23. D
24. D
25. D
26. B
27. A
28. B
29. C
30. A
31. B
32. A
33. B
34. A
35. D
36. B
37. A
38. B
39. D
40. B